Five Square Roots

Five
Square
Roots

the remarkable fecundity of
impossible square roots

Richard A.W. Bradford

principia publications unlimited

This edition first published March 2023

A catalogue record for this book is available from the British Library

Cover design by rawb

https://principiapublications.wordpress.com/

ISBN: 978-1-8380216-3-4

The Shorter Contents

The Longer Contents

0

The Root of this Book

Emulating E.T.Bell, I have cunningly tricked perennial preface-skippers by avoiding that term in favour of chapter 0 – at which, in any case, all books of mathematical leaning should really start.

The use of the word "root" in mathematics has come to us from the Arabic (jadhr), via Latin (radix), meaning root, as in plants, and hence by metaphor meaning the source or foundation or basis of something. In mathematics, "root" has also come to refer to the solution of an equation. But that is misleading. A polynomial can be fully reconstructed from its roots, that is, from the set of all values of its variable which make the polynomial equal to zero. So the polynomial itself can be considered as "springing from its roots". Hence it is the interpretation of "root" as a source or foundation that is apposite here, not really its meaning as the solution of an equation, which is derived from the former, more fundamental, meaning.

Thus, the root of this book lies in some particularly rich mathematical loam whence springs a sequence of mathematical structures of increasing power. These structures do indeed grow from roots – square roots, in fact.

The remarkable thing about all five instances of square roots addressed in this book is that they were all considered impossible or meaningless before they were conjured into existence by an act of cognitive will. Or perhaps should that be "before their existence was discovered by excavation of their chthonic lair?". You decide.

What is clear is the intellectual power we gain, whether it be by discovery or invention. The deductive power of the structures conjured into existence by these impossible square roots is as marked as it is baffling. And where mathematics leads, physics is not far behind. The more recent square roots are inseparable from their physical interpretations.

This is a maths book, a book about maths, and also in its later parts a physics book. But, undoubtedly because I am a physicist, even in the earlier parts it is the calculational power which the mathematics provides that interests me most, rather than mathematical rigour *per se*. These impossible square roots are astonishingly fecund, not just within mathematics itself but within physics too – and elsewhere.

This is not a book which obsesses about mathematical pedantry. Despite treating some rather "simple" subjects, it is not a book about foundational issues either, though we will be obliged to touch briefly on such matters in chapter 1. You can define the number "5" as the class of all sets which have a one-to-one (bijective) mapping with the set $\{1,2,3,4,5\}$ if you really want, but that is not the sort of thing we will be doing in this book – though it is pointed out that one has no choice when dealing with transfinite numbers. For one thing, once one starts heading 'downwards' in the direction of foundational issues, there is a tendency to vanish down an infinite rabbit hole (for example, down the rabbit hole of unresolved issues in axiomatic set theory).

In this book we will be heading in the opposite direction: the direction which is primarily about building ever richer mathematical structures. This is the direction most favoured by physicists (such as the author) who value mathematics for its power. This power involves the ability to express quantitative ideas both succinctly and precisely, as well as providing computational schemas to facilitate deductions and calculations. But it is more than that. These mathematical structures can be the source – the root, one might say – from which the physical concepts emerge.

Whence cometh, as Eugene Wigner had it, this "unreasonable effectiveness of mathematics in the natural sciences"? It is as mysterious as ever. Perhaps the origin lies in Leibniz's "pre-established harmony". The characteristic of this tantalising phenomenon is that, the more one studies it, the more mysterious it becomes. Here's Wigner,

"The miracle of the appropriateness of the language of mathematics for the formulation of the laws of physics is a wonderful gift which we neither understand nor deserve. We should be grateful for it and hope that it will remain valid in future research and that it will extend, for better or for worse, to our pleasure, even though perhaps also to our bafflement, to wide branches of learning."

The reader will rapidly discover that, in addressing the square roots which form the principal subject of each chapter, I take the opportunity to wax discursive on what springs from them. The subject matter of the book is not merely the roots, but the whole plant which grows therefrom. Well, not the whole plant, of course – that would hardly be possible. But sufficient, I hope, to give some indication of the unexpected luxuriance which unfolds from these roots.

It is possible that much of the book will appeal to a general reader without (say) A Level knowledge of maths, though chapter five will be an exception. Certainly, the book is not intended for the expert, who would find the contents elementary and also a rather random selection (being chosen largely by the author's fancy from a virtual infinity of options). The likely appreciative audience (if I may presume that to be a non-null set) will be keen A Level students of maths or physics, or undergraduates of these subjects. I flatter myself that even postgraduate physicists may find things here to entertain or enlighten.

Matters Presentational

Do use the Longer Contents to find things. This stands in lieu of an Index.

References are given at the end of each chapter, rather than all together at the end of the book. This is not supposed to be a work of great scholarship, but more an exercise in pedagogy. Consequently, I have kept the number of references down to just a few that happen to appeal to me. In practice, these days, readers will turn to internet searches if they want more detail on any topic.

Where I judge that a topic would unduly strain readers' patience, or is of greater difficulty, but I cannot resist its inclusion, I have relegated it to the Appendices, of which there are 12.

References

Bell, E.T. (1937), *Men Of Mathematics*. Simon & Schuster. 1987 edition now available as an eBook, https://www.simonandschuster.co.uk/books/Men-of-Mathematics/E-T-Bell/9780671628185

Stanford Encyclopedia of Philosophy. (2020). *Leibniz's Philosophy of Mind*. https://plato.stanford.edu/entries/leibniz-mind/

Wigner, E.P. (1960). *The Unreasonable Effectiveness of Mathematics in the Natural Sciences* in The Collected Works of Eugene Paul Wigner, Part B, Historical, Philosophical, and Socio-Political Papers, Ed. J.Mehra https://link.springer.com/content/pdf/10.1007%2F978-3-642-78374-6.pdf

1

The Square Root of 2

Wherein we find that, not only do irrational numbers exist, i.e., numbers which cannot be expressed as exact fractions, but that, in a strictly defined sense, almost all real numbers are irrational. We also meet algebraic numbers and their opposite, the transcendental numbers. We find that almost all real numbers are transcendental.

No attempt is made in this chapter to examine the motivations behind the historical development of numbers and their computation. In some cases, especially those historical individuals whose names we know from written works, the motive would appear to be purely, or mainly, intellectual. However, this belies the motives within the general population, whose keen interest would hardly have been drawn by anything so impractical. Famously, the ancient Egyptians' interest in geometry lay in applications to agriculture and building. The interest in numerical calculation in China, India, the Arab and Persian worlds, and later in the West, was driven by trade and banking. But let us not lose sight of the origin of the subject matter of this chapter, the irrational numbers, which arose in ancient Greece in the minds of those who were mainly philosophically motivated. Archimedes, for example, applied his knowledge to practical matters only with the greatest reluctance, despite their impressive, and sometimes devastating, successes. This interplay between the practical and the intellectually arcane runs through all the subjects of this book.

1.1 The Positive Integers

The origin of the positive integers in counting objects is clear and needs no explanation. However, the notation used to express them has a substantial history. Symbolism – that is, notation – matters. It facilitates or frustrates computation. Try doing multiplication with Roman numerals. There is a deep history associated with both notation and the closely associated issue of computational methods – including not just addition and subtraction but multiplication, division, raising to powers and, yes, the extraction of roots. However, those matters are not our concern here.

The main purpose of these introductory sections is to review how the concept of "number" developed, gradually admitting an ever-broader class of entity under that rubric. The first extensions beyond the positive integers were zero, fractions and negative numbers.

The integers less than one have their own history, to which it is worth alluding though we cannot dwell long upon it. They provide the first example of how extending a mathematical system into a regime initially thought trivial or meaningless actually increases the power of the system – and to an unreasonable degree.

Fractions provide an example in which notation plays a crucial role in facilitating, or frustrating, computation. The alternative notation of fractional quantities as decimals provides a different computational scheme to that of fractions: different and with some significant advantages, but by no means always superior. The major advantage of the decimal notation is conceptual. It paved the way for the entire gamut of real numbers – the continuum real line – complete with irrational numbers and transcendental numbers. The latter are the destination towards which we will work in this chapter.

1.2 Zero

There is a lot to say about zero. Many books have been written on the historical origins of zero, for example Seife (2000), Kaplan (2000) and Aczel (2015). Without doubt it was a major conceptual breakthrough to appreciate that zero was a number. Typical of mathematical progress, this seems trivial to us now. But the admission of zero to the pantheon of integers marked a move from counting to symbolic abstraction. Also typical of mathematical progress, it was found that acknowledging zero to be a number leads to unanticipated simplifications in later applications. How messy would the theory of the roots of polynomial equations be if zero was not a permitted solution? Indeed, one could not even define a polynomial equation by equating a polynomial to zero if the existence of zero was not already accepted.

But zero has also a second significance: as a placeholder in the Hindu-Arabic expressions of integers and decimals. This is equally foundational as it means that any integer or fraction can be expressed in terms of the ten digits with which we are now familiar.

The history of both uses of zero is complex, with various usages of greater or lesser sophistication occurring in Mesopotamia, India and China at various stages, some being several thousand years old, with more sophisticated usage appearing, especially in India, throughout the first millennium AD. Remarkably, the ten-digit decimal system of numerals was introduced into the West only in 1202 by Fibonacci. They became known to us as Arabic numerals because their transmission from India was via the Arabic (and Persian) world. They are now known more correctly as the Hindu-Arabic system. This decimal system, with its crucial zero digit, hugely facilitates practical computation.

1.3 Fractions

Contrary to popular belief, fractions were not invented specifically for the purpose of torturing school children. (Remarkably, Latin was not invented for that purpose either, though I find that harder to believe).

By "fraction" I refer to the notation of a type of number expressed as a ratio of two integers. A fraction, in this sense, is not necessarily less than 1. If it is, it is said to be a "proper fraction". Conversely, an improper fraction is greater than or equal to one.

The origin of the idea of fractions is obvious and does not need labouring. One does not need to be a mathematical genius to appreciate that a piece of cake is only a fraction of the whole cake. Whether one uses the word "fraction" or not is irrelevant; the concept is clear and self-evident. However, notation and computational methods are another matter.

There is, of course, no shortage of nonsense on the internet. One such is the claim that *"fractions as we use them today didn't exist in Europe until the 17th century"* (Pumfrey, 2022). This would imply that fractions were unknown to Fibonacci, Tartaglia and Copernicus – all of whom used fractions as routine.

Fractions in a form close to what we use today were certainly in use in India in the first millennium AD, and with increasing sophistication in the second half of that millennium. The Jain mathematician Mahāvīra is perhaps most noteworthy as his written record contains extensive instructions for carrying out computations with fractions (circa 850 AD). He had a sequence of Indian precursors, such as the Hindu Brahmagupta (circa 598–670 AD), who, incidentally, was amongst the first to involve zero in computational schemes. The Indian notation for fractions was essentially the same as ours, though without the horizontal bar. Also, improper fractions were written with the integer part first, so that they would write,

$$2\frac{3}{5} \quad \text{as} \quad \begin{array}{c} 2 \\ 3 \\ 5 \end{array}$$

It was the Arabs (or Persians) who reputedly introduced the horizontal bar into the notation. The first use of it in the West is due to Fibonacci in 1202 in his *Liber Abaci*. This work also introduced the Hindu-Arabic decimal notation to the West, though apparently it was slow to catch on amongst the merchant class (the astronomers would be a different matter). The advantages of decimals over fractions when performing certain calculations began to be more widely appreciated only after the publication of an influential book by the Flemish mathematician Simon Stevin in 1585. So, contrary to the above preposterous claim that fractions became used in the West only in the 17th century, actually by then fractions were already "old hat" by hundreds of years.

Since this first chapter will largely be about the Pythagoreans' shock at discovering that some numbers could not be expressed by fractions at all, it is worth spending a little time singing the praises of fractions. Sometimes fractions can be more convenient than decimals, especially if you want exactitude. Decimals, one soon discovers, have a habit of being infinitely long. Providing that one confines attention to rational numbers – that is, numbers which can be expressed as a fraction – there is a decimal notation which makes the infinite sequence of digits clear, namely the "recurring" notation, e.g., $(abc)' \equiv abcabcabc$ ..... For example,

$$\frac{1}{3} = 0.3333 \ldots = 0.(3)'$$

$$\frac{1}{7} = 0.(142857)'$$

One can hardly dispute that the notation as a fraction is better in the second example. Similarly, precise computations can be better done in fractions, for example,

$$\frac{7}{13} \times \frac{5}{11} = \frac{35}{143}$$

using the rule,

$$\frac{a}{b} \times \frac{c}{d} = \frac{a \times c}{b \times d}$$

In decimal notation that very easy calculation would be rather a pain to work out via long-multiplication,

$$0.(538461)' \times 0.(45)' = 0.(244755)'$$

This is a fact from which we are now insulated by calculators – providing we are content with limited precision.

Just to drive home the relative compactness of fractions if one seeks precision, consider the reciprocals of prime numbers. The decimal expansion of the reciprocal of an integer, prime or not, will always either terminate or involve recurrence, but the number of digits before recurrence occurs may be huge. In fact, in the worst case, the reciprocal of a prime p can involve recurrence of a sequence of up to $p - 1$ digits. $p = 7$ is an example, and so is $p = 60{,}017$. The reciprocal of the latter can be written as a decimal with full precision only by writing down a sequence of 60,016 digits. Good luck with that. I think you will agree that 1/60,017 is more compact and more readily digested by one's intuition.

One of the first programs I ever wrote, back in 1972, was a suite of routines to carry out computer calculations using exact fractions, rather than decimals of necessarily limited precision. There was a purpose: I was using an iterated procedure to solve some equations in which the rounding error associated with "real" (i.e., decimal) numbers accumulated to point of overflow (divergence). Shifting to fractions proved successful. Though perhaps not the optimal solution in coding terms, it proved a point which I have never forgotten.

If you wish to provoke students into thinking (dreadfully out of fashion, I know) ask them to judge the truth or falsity of,

$$1 = 0.(9)'$$

(True, btw).

1.4 Fractions, Ambiguity, and Prime Factorisation

There is a bit of a problem with fractional notation which we glossed over, but which was addressed by the aforementioned Indian mathematicians: fractions which look different may be the same. For example,

$$\frac{65}{1183} = \frac{5}{91}$$

The reason is that any common factor shared by the numerator (the top number) and the denominator (the bottom number) can be cancelled from both. This follows from the multiplication rule already given. In this example,

$$\frac{65}{1183} = \frac{5 \times 13}{91 \times 13} = \frac{5}{91} \times \frac{13}{13} = \frac{5}{91} \times 1 = \frac{5}{91}$$

A prime number is any positive integer which has exactly two distinct divisors, namely itself and 1. Note that 1, therefore, is not prime because of the requirement for exactly two distinct divisors. The Fundamental Theorem of Arithmetic states that every integer greater than 1 which is not prime can be expressed as a product of two or more primes in a unique manner. (Note that the inclusion of 1 as a prime would wreck that definition).

The Fundamental Theorem of Arithmetic saves fractional notation from ambiguity because it shows that every fraction can be expressed in a unique "lowest form". This lowest form is obtained when every common prime factor is cancelled between numerator and denominator. In the above example the prime factorisations are,

$$\frac{65}{1183} = \frac{5 \times 13}{7 \times 13 \times 13} = \frac{5}{7 \times 13} = \frac{5}{91}$$

So there was only one common prime, i.e., one factor of 13. Another example with two common prime factors is,

$$\frac{630}{1029} = \frac{2 \times 3^2 \times 5 \times 7}{3 \times 7^3} = \frac{2 \times 3 \times 5}{7^2} = \frac{30}{49}$$

This is one way in which the decimal notation scores over fractions: in decimal notation two numbers are equal iff every digit is equal – with the one exception of recurring 9s.

(NB: The abbreviation "iff" stands for "if and only if").

1.5 Proof of the Fundamental Theorem of Arithmetic

That every positive integer can be factorised into a unique product of primes was recognised and proved by Euclid circa 300 BC (albeit expressed differently). It can be proved in many different ways. Here's a proof by *Reductio ad Absurdum*.

Suppose that one or more non-prime integers greater than 1 can be expressed as a product of primes in two different ways. There is therefore a smallest such integer which we call X. We can write its alternative factorisations as,

$$X = p_1 p_2 \ldots p_n \text{ and } X = q_1 q_2 \ldots q_m$$

It immediately follows that all the prime factors must be different because, if two prime factors were equal (which we may take to be $p_1 = q_1$ without loss of generality) then at least some of the others must differ, by assumption, so that the integer $\dfrac{X}{p_1}$ would equal $\dfrac{X}{q_1}$ and hence be expressible as two different products of primes. Since this integer is clearly smaller than X this contradicts the initial assumption that X is the smallest such integer.

Consequently we may now assume $p_1 > q_1$ without loss of generality and write $X = p_1 P = q_1 Q$ and $(p_1 - q_1)Q > 0$ by assumption so that,

$$(p_1 - q_1)Q = p_1 Q - q_1 Q = p_1 Q - p_1 P = p_1 (Q - P) > 0$$

But Q is clearly less than X and hence has a unique prime factorisation, by assumption, namely $Q = q_2 \ldots q_m$ and all these factors differ from p_1. Hence the above expression can hold only if $(p_1 - q_1)$ were divisible by p_1 which it clearly is not as q_1 cannot be zero and so $(p_1 - q_1)$ must be less than p_1. This establishes a contradiction and hence there is no non-prime integer greater than 1 expressible as a product of primes in two different ways. QED.

In the next chapter we will see that proving a parallel theorem about factorising, namely the Fundamental Theorem of Algebra, is a good deal more subtle.

1.6 The Square Root of 2

Pythagoras of Samos (circa 570 to 495 BC) was a somewhat mysterious figure, it being tricky to distinguish his own work from that of the Pythagorean school he founded. Like Euclid, despite being Greek, his main work was carried out elsewhere than in Greece. In Pythagoras's case in southern Italy (in Euclid's case, in Alexandria, Egypt). Pythagoras is said to have first recognised what we now know as the five Platonic solids. He taught metempsychosis, the immortal nature of the soul and its transmigration into a new body on death. If this puts you in mind of Buddhist teaching, note that the Buddha (Siddhartha Gautama) was closely contemporaneous with Pythagoras, possibly circa 563 to 483 BC. (To avoid upsetting Buddhists I note that the idea of an immortal soul is anathema in Buddhist teaching whose doctrine is centred on the Middle Way between eternalism and nihilism).

We'll meet the Pythagorean musical scale shortly. Pythagoras is credited with the idea of "the music of the spheres", i.e., that the motion of the planets is analogous to music in respect of being based on harmonious simple ratios. Completely wrong for the planets, of course, but a jolly good effort for musical scales, though with a bit of a problem, as we shall see later.

Poor Pythagoras's faith in exact fractions was trashed by his own discovery, closely related to his most famous work: Pythagoras's Theorem. The latter relates the length of the longest side of any right-angled triangle (the hypotenuse) to the lengths of the other two sides: the square of the hypotenuse equals the sum of the squares of the other two sides. It is believed that the Pythagorean theorem was known and used by the Babylonians and Indians centuries before Pythagoras. The simplest example is illustrated by Figure 1.1.

If each of the smaller green squares has side 1, then the area of each is also 1. The blue square formed from their diagonals clearly has twice this area, because it consists of four triangles whereas the green squares consist of just two. So, if the area of the blue square is 2, its sides must be of a length such that, when squared, they give 2, i.e., their length is root 2, or $\sqrt{2}$.

Figure 1.1: Pythagoras's unwanted discovery

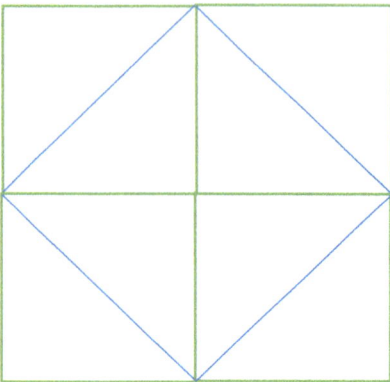

Naturally, having come to this clear and unambiguous conclusion, Pythagoras, with his obsession with fractions, would have sought the fraction which equals this $\sqrt{2}$. Whether it was Pythagoras himself, or one of his school (perhaps Hippasus of Metapontum) who first realised that this was impossible is not clear. Once again it has been claimed that the Indian mathematicians got there first, but this is disputed.

The proof is extremely simple assuming that the Fundamental Theorem of Arithmetic has already been established. Again we proceed by *Reductio ad Absurdum*. Suppose there are integers a and b such that $\frac{a}{b} = \sqrt{2}$. The Fundamental Theorem of Arithmetic allows us to assume that this has been put in its lowest form, so that a and b have no common factors. It immediately follows that $a^2 = 2b^2$ in which case a must be even and hence can be written $a = 2c$ for an integer c. In which case $2c^2 = b^2$ so that b is also even, and thus has a common factor with a, namely 2. But this contradicts the assumption of lowest form, and hence no integers a and b such that $\frac{a}{b} = \sqrt{2}$ exist. QED.

So here we have the invention, or discovery, of irrational numbers – numbers which cannot be expressed as a fraction. Note that the demonstration based on Figure 1.1 is essentially geometrical; it relies on the concept of area, and upon the area of a square being the product (hence the "square") of its sides.

Already we have used *Reductio ad Absurdum* twice. We will see in chapter 4 that the validity of this familiar workhorse of mathematics can be challenged.

1.7 Pythagoras and Musical Scales

Since Pythagoras is by way of being the hero of this chapter, it is apposite to point out another area in which his insistence upon exact fractions, in contrast to the real number continuum, was ultimately to prove unworkable – though it was darned good effort. I refer to musical scales – or, to be more specific, the Western chromatic scales. For details I refer the reader to Bradford (2009). In brief, the Pythagoreans' insistence on simple ratios is what led to the Western chromatic scales consisting of eight notes and 12 semitones. But the fly in the Pythagorean ointment is the wolf note or Pythagorean comma. The awkward fact of the Pythagorean scale based on simple fractional ratios is that the octave note fails to be at exactly twice the tonic, and hence fails to be truly harmonious. It is a frustratingly close miss at,

$$\frac{3^{12}}{2^{18}} = 2.02728653 \ldots$$

Another practical problem with the Pythagorean scale, even within the first octave, is that the semitones are not equally spaced (in frequency-ratio terms) but have two different spacings. Guitars could not then have frets spanning the whole neck (as if left-hand fingering were not problematic enough already) and would need to have fixed tuning. Out would go guitarists' love of unconventional tunings as well as barred chords.

These practical problems are all eliminated at a stroke by adoption of the equal-tempered scale, in which all 12 semitones are equally spaced (in ratio, or log(frequency), terms). But by adopting the equal-tempered scale, which became universal around the time of J.S.Bach, musicians implicitly abandoned Pythagorean exact fractions in favour of the decimal world and irrational numbers.

Thus, the adoption of the equal-tempered scale is Western musicians' equivalent of mathematicians adopting irrational numbers. And in both cases it leads to simplifications in what follows thereafter.

1.8 Irrational Numbers

A brief note on etymology: it is tempting to regard the origin of the use of the term "irrational" to denote numbers like $\sqrt{2}$ as lying in the lack of an expression of them as ratios, hence "irrational" equates to "not ratio". One might then dismiss the idea that the terminology suggests that such numbers are contrary to reason, that is "irrational" in the everyday sense. But this comforting etymology would seem to be false. Euclid called irrational numbers "alogos", which does indeed mean contrary to logic, or unreasonable.

Whatever the etymological chronology, the terminology does betray the shock the ancients experienced in discovering the existence of such numbers which, to their minds, were contrary to reason. It is a peculiar point of view, however, in that these numbers are actually conjured into existence purely by reason. A better term might be "unexpected" numbers. Or, better still, and in spite of the true etymology, let us cling to the interpretation of "irrational" as meaning the lack of expression as a ratio – which has the merit of being the correct definition.

Just as the expression of a number as a ratio fails for the irrationals, so does the expression as a decimal. The decimal expansion of an irrational number involves an infinite sequence of digits which never repeats, and hence the "recurring" notation fails to save us. An irrational cannot be exactly specified as a decimal with a finite number of digits any more than it can by a ratio of integers (as a moment's reflection makes clear).

Armed with the Fundamental Theorem of Arithmetic it is simple to show that the square root of any positive integer is either an integer or irrational, i.e., a number which is not a perfect square always has an irrational square root.

The numbers π and e are irrational (very irrational, one might say – they are transcendental – see below).

So irrational numbers are not rare, then. Oh, but that is a serious understatement. In a strictly well-defined sense, "almost all" real numbers are irrational.

In order to explain what that means it is necessary to define what is meant by "all numbers". I refer here only to real numbers, not imaginary or complex numbers (for the excellent reason that we have not yet brought these into existence and will not do so until chapter 2). However, I do include negative numbers about which I shall say a few words shortly. Here I encounter a difficulty with precision if I wish to avoid going down a mathematical rabbit hole. So, if you really want to see the definition of the real numbers as the unique Dedekind-complete ordered field up to isomorphism, then I suggest you look it up (don't all rush at once).

For our present purposes I shall be content with the rather imprecise definition of the reals as analogous to points on a linear continuum (a line), or, if you prefer, a real number is a decimal expression, extending potentially to an arbitrarily large number of digits. For the irrationals we must, of course, admit an infinite number of digits.

In what sense, then, are "almost all" real numbers irrational? The key to understanding this is to appreciate that, whilst there are an infinite number of rational numbers there are also an infinite number of real numbers, and the latter infinity is infinitely bigger than the former infinity! Yes, that statement can be made mathematically well defined, as follows.

Figure 1.2: The countability of the rational numbers

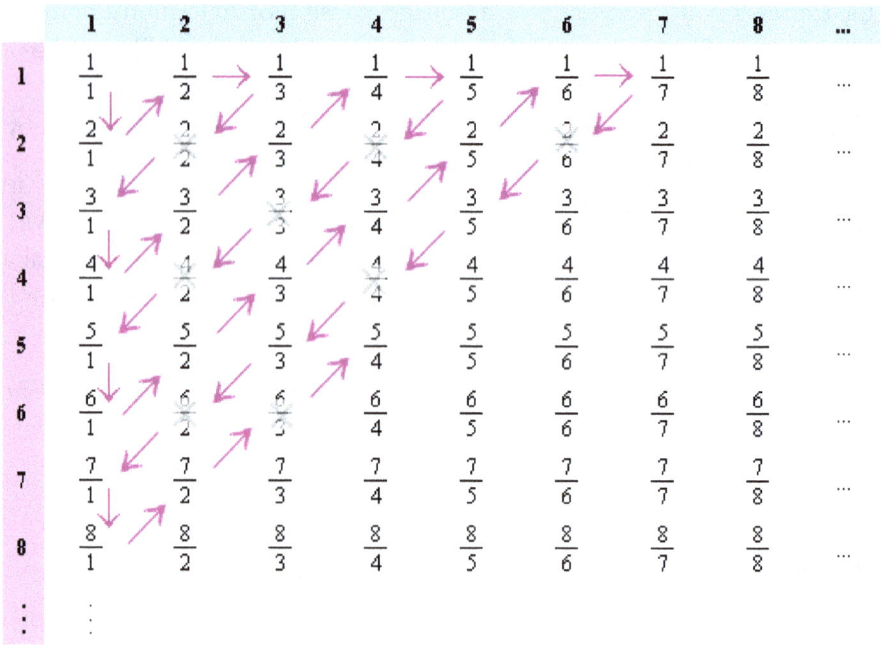

A countable set is a set whose elements can be counted! That means one can define a one-to-one (bijective) mapping between the set and the set of positive integers. This can easily be done for the rational numbers. For example, we could simply write down all the rational numbers in a square grid, with the numerator defining the row and the denominator defining the column, as in Figure 1.2. We can then assign a unique single integer index to every fraction (proper or improper). Hence, whilst there is an infinite number of rational numbers, this infinity is the same infinity as the number of integers. That these two infinities are equal is established by both being countable, and hence in one-to-one correspondence with each other. This infinity is called aleph-null and denoted $\aleph_0$.

It should be clear to the reader that all integers, including the negative integers, are countable and hence that there are "the same number" of integers as positive integers, both being the same as "the number of" rational numbers, namely $\aleph_0$.

But what happens if we try to do something similar for the real numbers? The proof that the real numbers are not countable proceeds by *Reductio ad Absurdum* yet again. Assume the reals are countable and write down the reals in (infinite) decimal form in the order they map to the positive integers. (For this purpose it does not matter if you use base 10 numbers, or binary numbers, or any other number base). You then define a real number as follows,

- Define its first digit as any digit not equal to the first digit of the first number;
- Define its second digit as any digit not equal to the second digit of the second number;
- Define its third digit as any digit not equal to the third digit of the third number;
- And so on, for an infinite number of digits.

It is clear that the number constructed must differ from every number in the list, because, by construction, it differs from every number in the list by at least one digit. Consequently the initial list was incomplete as it did not provide an integer index for this number. It does no good to add this missing number – say at the beginning of the list – since the construction applied again will always produce another missed number. Hence, the real numbers are not countable.

The "number of real numbers" is denoted $\aleph_1$ (aleph-one). It is a bigger infinity than $\aleph_0$. (The reader will appreciate that I'm taking liberties with mathematical rigour here).

Note that the same argument can be applied to any finite interval of the real line, say $[a, b]$, where $a < b$, thus indicating that the number of numbers in any finite interval is also $\aleph_1$. In geometrical terms, the number of points on any finite line segment is indenumerably infinite, i.e., $\aleph_1$.

If $\aleph_1$ is the "number of points along a line" you may be tempted, as I was, to make the error of assuming that $\aleph_2$ is the number of points in a two-dimensional area, and similarly that $\aleph_3$ is the number of points in a three-dimensional volume, etc. This is wrong. Actually, the "number of points" (properly, the "cardinality") in any N-dimensional region is $\aleph_1$, independent of N (assuming finite N).

It is very simple to establish this. The N-dimensional region is defined by N real numbers. We need to establish an injective map onto the reals, i.e., to assign a unique real number to each set of N real numbers. This can be done, for example, by using the first digits of each of the N numbers to be the first N digits of the unique number. The second digits of the N numbers then define digits N+1 through 2N of the unique number, and so on. This number then uniquely specifies a point in the N-dimensional space because all N 'coordinates' can be reconstructed from this single real number.

This gives you some feel for how humongously large $\aleph_2$ must be.

These ideas were first made rigorous by Georg Cantor in his work on transfinite numbers. Introducing rigour takes us into deep waters. Let us approach with some caution.

1.9 Almost All Real Numbers are Irrational

We have already seen that the number of real numbers ($\aleph_1$) is greater than the number of rational numbers ($\aleph_0$), and we can assert this despite both those "numbers" being infinite. That imprecise statement can be made rigorous by the replacement of "the number of" items in a set with the concept of cardinality, defined by the sort of injective mappings described in the previous section.

Continuing to proceed in an obviously questionable but intuitively appealing way, then if the number of real numbers is $\aleph_1$ and the number of rational numbers is $\aleph_0$ then the number of irrational numbers is surely $\aleph_1 - \aleph_0$. But the size of the infinity represented by aleph-one is so stupendously larger than the mere countable infinity represented by aleph-null that $\aleph_1 - \aleph_0 = \aleph_1$. In other words, vanishingly few real numbers are rational and so "almost all" real numbers are irrational.

I will hedge this around with a host of provisos shortly, but the relationship between $\aleph_0$ and $\aleph_1$ can be written (professional mathematicians should look away at this point to avoid offence) $\aleph_1 = 2^{\aleph_0}$.

To drive home what this relationship implies, it is a useful exercise for the reader to evaluate what fraction $\aleph_0$ would be of $\aleph_1$ if they were assumed finite but as ever larger values for $\aleph_0$ were considered. We get the following,

$\aleph_0$	$\aleph_1/\aleph_0$
10	$\sim 10^3$
100	$\sim 10^{28}$
200	$\sim 10^{58}$
300	$\sim 10^{88}$

You get the picture. Even at a paltry $\aleph_0 = 300$, $\aleph_1$ is already 88 orders of magnitude larger than $\aleph_0$. In the limit of infinite $\aleph_0$ it becomes a fraction of $\aleph_1$ which is "exactly zero".

In this sense, the rational numbers are so sparse within the set of all real numbers that the fraction of real numbers which are rational is zero. If you choose a real number at random, it is certain to be irrational. You can envisage this by building a random real number by randomly assigning each digit, one at a time. The probability that any digit is 0, assuming base 10, is 0.1. The probability that a sequence of n consecutive digits are all zero is 10^{-n}, and this heads rapidly to zero for unbounded n. But any rational number whose decimal form truncates would end with an infinite sequence of zeros, and so has "zero probability" of resulting from this process. For a rational number whose decimal expansion has a recurring sequence the situation is actually just the same. Every new digit you include is known from the recurring pattern, and so also has a probability of 0.1 per digit of occurring by random chance.

Now I must confess my many sins (how long have you got?). Focussing on the expression $\aleph_1 = 2^{\aleph_0}$ two things must be made clear. Firstly, it is not necessarily correct, and secondly, it doesn't mean what it seems to mean. The second of these is the easier to address.

Put on your lifejacket, we're going into deep waters.

When dealing with transfinite numbers – which is what we are truly about here – it is always really the cardinality (i.e., the mapping properties) of the sets in question which we should consider. In set theory (at least, the sort of set theory introduced below) there is a concept of a "power set". The power set of a set S is denoted 2^S. It is defined as the set of all mappings from S to a two-element set, say $\{0, 1\}$. This is well defined even for infinite sets. Note that the power set is another way of talking about how many different ways there are of partitioning S into two parts. It is clear that if S were a finite set with N elements, then this number would be 2^N, hence the notation. So, strictly, in the notation $\aleph_1 = 2^{\aleph_0}$ we are referring, not to numbers but to the sets whose cardinalities are $\aleph_0$ and $\aleph_1$.

Now for the thornier issue that the expression $\aleph_1 = 2^{\aleph_0}$ is not necessarily correct.

This question is closely related to an hypothesis first made by Cantor, known as the Continuum Hypothesis. It claims that there is no set whose cardinality is strictly between that of the integers ($\aleph_0$) and that of the real numbers ($\aleph_1$). The Continuum Hypothesis claims that, in the world of transfinite numbers, there are none between $\aleph_0$ and $\aleph_1$. Cantor believed it to be true but failed to prove it. To do so was one of David Hilbert's famous set of 23 mathematical challenges he posed at the start of the twentieth century. At the time of writing, eight of those problems have been solved, and six have definitely not been solved. A further nine have partial solutions, or possible solutions which are contentious in some respect. The Continuum Hypothesis was problem number one! It is one of those which falls in the "partial solution" camp.

The provability of the hypothesis depends upon how you formulate your theory of sets. A purely informal set theory runs into problems like the Russell paradox – the set of all sets which are not members of themselves. It is a self-denying entity. The purpose of axiomatic set theories is to avoid such issues, and to establish agreed sound principles of reasoning. The *de facto* standard axiomatic set theory has become that denoted ZFC, which stands for "Zermelo–Fraenkel set theory with the axiom of choice". No, I'm not going into what that is – interested readers can look it up. The important thing is that, within ZFC set theory, the Continuum Hypothesis is equivalent to the statement $\aleph_1 = 2^{\aleph_0}$.

So far, so good. That means, if we are willing to accept ZFC set theory as the basis of mathematics, then the rather arcane power-set claim that $\aleph_1 = 2^{\aleph_0}$ has the more graspable interpretation that there is no set whose cardinality is strictly between $\aleph_0$ and $\aleph_1$.

It has been shown that the Continuum Hypothesis is neither provable within ZFC set theory, nor can it be disproved within that system. This is known as the "independence of the continuum hypothesis", meaning its truth or falsity is independent from the axioms of ZFC set theory.

Where one goes from there is problematic. One option, associated with the formalist school of mathematical philosophy, is that we are free, if we wish, to add the Continuum Hypothesis to the axioms of ZFC set theory to provide an enriched set theory (though one could equally well add the falsity of the Continuum Hypothesis as a new axiom instead, and the two opposing set theories would, one presumes, then present us with two incompatible mathematical universes – an uncomfortable notion).

The other option, associated with the Platonist school of mathematical philosophy, is that the Continuum Hypothesis is either true or false and that this is an absolute fact, one way or the other, independent of puny human reasoning. Gödel, for example, was of this opinion. He also believed the Continuum Hypothesis to be false. He was the first to show that ZFC set theory is compatible with the Continuum Hypothesis – which perhaps disappointed him – but his spin on that finding would be that it merely showed that ZFC set theory was inadequate.

Well, I did warn you that these were deep waters.

Let me close this section with another way of demonstrating that "almost all" real numbers are irrational, and one based on analysis rather than set theory. Consider a function $D(x)$ of a real number, x, defined as 1 when x is rational but 0 when x is irrational (known as the Dirichlet function). Then consider the integral of this function, say between 0 and 1, though it could be over any real interval,

$$\int_0^1 D(x)d\mu(x) = 0$$

I have given you the punchline: the integral of the Dirichlet function over any real interval is zero. This is another way of envisaging the extreme sparsity of rational numbers within the real line. Choose x at random and it's certain to be irrational, so $D(x)$ is certain to be zero.

The alert reader will have spotted two things: firstly that the usual Riemann integral is surely not well defined for such a function, and secondly that the "measure interval" $d\mu(x)$ appears above, rather than the usual dx. The integral above must be understood as a Lebesgue integral, not a Riemann integral. And so we come full circle, because the Lebesgue measure $d\mu(x)$ can be found only on the basis already discussed, namely from set-theoretical considerations, injective mappings, countability, etc. Nevertheless, the formulation in a manner which looks like conventional analysis is comforting to some of us pedestrians.

1.10 Negative Numbers

One can think of two apples or three apples, but it is rather tricky to think of minus two apples. However, in the world of trade and banking it becomes more natural under the concept of "being owed". Book-keeping, in the accountancy sense, inevitably involves both credits and debits. And accountants must necessarily keep tabs on what they are owed, and what others owe to them. The concept of having an overdraft with your bank, being in debt on your credit card, or having other outstanding debts, these are familiar to us all (unfortunately). How much money do you then have? Clearly not just none, but less than none.

I have motivated irrational numbers here as they arose in the minds of the ancients, through geometry. But there is another way: through analysis – that is, the mathematics of functions and equations. If we did not allow irrational numbers than an equation as simple as,

$$x^2 - 2 = 0$$

would have no solution. The same approach also motivates negative numbers. An even simpler equation, such as,

$$x + 2 = 0$$

would have no solution if we did not allow negative numbers. Time and time again we will see this phenomenon: the scope of what is meant by "number" expands to permit solutions to equations which would otherwise not exist.

We can now talk of integers meaning either positive or negative integers (or zero). And we can talk of rational and irrational numbers, both of which will include those which are negative as well as positive (or zero). The continuous real line now stretches from $-\infty$ to $+\infty$.

Move along, then. Nothing more to be seen here.

Oh, how wrong.

1.11 Algebraic Numbers

Mathematicians soon became interested in more complicated equations of the above type: polynomial equations. In the medieval period, Western mathematicians would set challenges for each other to find solutions, such as the famous controversy between Tartaglia and Cardano over the cubic equation, see Katz (1998). The most general polynomial equation order n is (for a positive integer n),

$$a_n x^n + a_{n-1} x^{n-1} + \cdots + a_1 x + a_0 = 0$$

We will meet these polynomial equations again in chapter 2. They have a rich history as mathematicians sought and found general solutions "in closed form" for the quadratic ($n = 2$), the cubic ($n = 3$), then the quartic ($n = 4$), but sought a general solution to the quintic ($n = 5$) in vain – and eventually proved that no closed form solution in terms of surds exists, again see Katz (1998). Do not confuse this with there being no solution, though, which would be a very different claim – and wrong, as it happens.

[Note that the term "surd" refers to an expression involving a square root, $\sqrt{}$, a cube root, $\sqrt[3]{}$, a fourth root, $\sqrt[4]{}$, etc.].

An "algebraic number" is any real number which can be expressed as the solution to a polynomial equation whose coefficients, $a_0, a_1 \ldots a_n$ are all rational (but may be positive or negative or zero).

It should be immediately clear to the reader than this is equivalent to a definition which insists that the coefficients $a_0, a_1 \ldots a_n$ are all integers (positive, negative or zero). To see this, multiply the equation with rational coefficients by the product of their denominators.

It is also clear that, (i) all rational numbers are algebraic, and, (ii) some irrational numbers are algebraic, e.g., every non-integer which is the nth root of an integer. That all rational numbers, say $x = p/q$, are algebraic follows from the obvious fact that they are solutions of the first order polynomial, $qx - p = 0$. Hence you can think of algebraic numbers as an extension of rational numbers into roots of polynomials with integer coefficients.

Real numbers which are not algebraic are called "transcendental numbers". All transcendental numbers are irrational, but not all irrational numbers are transcendental.

In Appendices B and C I give proofs that e and π are irrational. But are they also algebraic? They are not.

Proofs that e and π are transcendental are given in Appendices D and E. None of these proofs is trivial, far from it. They are not the sort of thing that your average pedestrian (*comme moi*) would produce unaided.

1.12 Almost All Real Numbers are Transcendental

Readers should now be asking themselves "how many" algebraic and transcendental numbers there are – and should now appreciate that the rigorous way of asking such questions – given that we are dealing with transfinite numbers of numbers – is via injective mappings. Specifically, are "almost all" irrational numbers (or reals) algebraic? Or, alternatively, are "almost all" irrational numbers (or reals) transcendental? The answers are "no" to the former and "yes" to the latter.

The reader should have a fair chance now of proving this. Have a go!

One way of proving that transcendental numbers exist, without actually constructing one, is Cantor's argument. It also demonstrates that "almost all" real numbers are transcendental. This starts with the observation that the reals are not countable and proceeds by showing that the algebraic numbers **are** countable, and hence that – not only must transcendental numbers exist but that (heuristically speaking) the number of them is aleph one minus aleph null, $\aleph_1 - \aleph_0 = \aleph_1$.

Here's a proof (which is mine and differs from Cantor's). We wish to establish that the algebraic numbers are countable, and hence that "the number of algebraic numbers" (i.e., their cardinality) is $\aleph_0$. If we can show this then it will follow that "almost all" real numbers are not algebraic, i.e., that almost all real numbers are transcendental. Now, by definition, every algebraic number can be defined by the set of coefficients $a_0, a_1 \ldots a_n$ in the polynomial equation which they obey, and these coefficients can be taken to be integers. For any such set of integers we can define a single integer by,

$$p_1{}^{a_0} p_2{}^{a_1} p_3{}^{a_2} \ldots p_{n+1}{}^{a_n}$$

where p_i is the ith prime number. This number is clearly different for every set of coefficients (considered, note, as an ordered set). Note that this follows from the Fundamental Theorem of Arithmetic. This establishes the required injective mapping, so the set of all algebraic numbers is countable.

The alert reader will spot a flaw in that proof because the set of coefficients $a_0, a_1 \ldots a_n$ will, in general, correspond to an equation with more than one solution – up to n distinct solutions. But this is not fatal to the proof. One merely has to define a further integer, r, by putting all the roots of the polynomial equation in some convenient order and identifying the required root as the rth. Then we add a further prime factor $p_{n+2}{}^r$ to the expression above which defines the mapped-to integer.

QED – almost. But even now there is a flaw in that we have assumed a result we are not yet entitled to assume, namely that an nth order polynomial equation has n roots. Actually that assumption could be replaced merely by the assumption that every nth order polynomial equation has a finite number of roots without the above argument being otherwise changed. Alternatively we could wait until, after chapter 2, we have proved that every nth order polynomial equation has not more than n real roots, at which point the proof is valid as it stands: the set of all algebraic numbers is countable.

Hence, "almost all" the real numbers are transcendental.

Despite that fact, it is generally very challenging to prove, of any given number, that it is indeed transcendental. The proofs that π and e are transcendental are given in Appendices D and E.

e^π is also transcendental, as is e^a for any non-zero algebraic number a. Also, a^b is transcendental for any algebraic a not 0 or 1 if b is any irrational algebraic number. So, for example, $2^{\sqrt{2}}$ is transcendental even though neither 2 nor $\sqrt{2}$ are transcendental.

However, it is not known (at the time of writing) whether $\pi + e$ or πe are transcendental (though at least one of them must be). It is also unknown at present whether e^e or π^π or π^e are transcendental (though I would eat my boots if any of those were not).

1.13 The Impossibility of Constructing the Angular Trisector

This topic provides a nice illustration of how far from the root the plant may grow, and yet still be entirely dependent upon it. The ancient Greeks enjoyed the challenge of constructing geometrical objects using only a straight edge and a pair of compasses. (Incidentally, I note that in place of "a pair of compasses" it is now common to see sources simply using the word "compass". In my day this would have resulted in derision from one's fellow pupils and a reprimand from the teacher. A compass is, of course, a magnetic needle which points north.)

Trisecting an arbitrary angle defeated the Greeks. They must have suspected it was impossible. But how does one prove the impossibility of such a thing? One can try forever and get no closer to proving impossibility. You could, in principle (if you were completely barmy) try to construct the regular polygon with 2048 sides – or that with 2047 sides. I doubt you would succeed, but if you were tempted to conclude both were impossible you would be wrong. On the other hand, if you actually succeeded in constructing the 2048-gon (and in principle you might) then you might be tempted to believe the 2047-gon was also constructible and go on trying until hell froze over. (It is impossible).

The secret is to translate the problem into the solution of polynomial equations and the types of numbers defined by their roots. We have defined algebraic numbers as numbers which are a root of a polynomial equation with rational coefficients. But algebraic numbers can be further subdivided according to the order of the polynomial of which the number is a root (called the degree of the algebraic number). This turns out to be the key to unlocking what was, in the early nineteenth century, an entirely new branch of mathematics: field theory, group theory and ultimately Galois Theory. In Appendix F I outline how the ideas of field theory are developed and lead to a proof of the impossibility of constructing the trisector.

The first published proof of the impossibility was due to Pierre Wantzel in 1837. He used the method of fields, essentially as I present in Appendix F.

However, the mathematical insights of Évariste Galois were of far greater importance than Wantzel's one-off proof. In 1830, at the age of 18, Galois had published a paper defining the finite field, another on the roots of equations, and a third outlining what we now call Galois Theory. Galois spent the night before his death compiling some notes, annotating an original manuscript and writing a letter describing his ideas. The next day, 30 May 1832 he was shot in duel and died the following day, aged 20. While alive his work had suffered rejection by Cauchy, failure to be understood by Poisson, and the misfortune of being sent to Fourier to review who unfortunately died before addressing it. It was 11 years after his death that the material which Galois had bequeathed the world on his last night was finally declared sound, by Joseph Liouville, and was published in 1846. Today, the mathematical topics outlined in Appendix F are subsumed in

Galois Theory, and the proof of the impossibility of trisecting the angle is merely a simple example of its power.

As an amusing finale, if any smart alec, knowing it to be impossible, bets you cannot trisect an arbitrary angle you might like to take up the challenge anyway. You can then confound him by stating that, for your construction, you will not need the rule, you will not need the pair of compasses either, and you will not even need the pencil! You may then proceed to construct the trisector using only origami paper folding, see the YouTube video by Zsuzsanna Dancso.

1.14 References

Aczel, A. D. (2015). *Finding Zero: A Mathematician's Odyssey to Uncover the Origins of Numbers*. St Martin's Press.

Bradford, R.A.W. (2009). *Musical Scales*. http://rickbradford.co.uk/MusicalScales.pdf

Cantor, Georg (1955). *Contributions to the Founding of the Theory of Transfinite Numbers*. Philip Jourdain (ed.). New York: Dover, NY. (English translation originally published 1915, Open Court Publishing).

Dancso, Z. (2014). *Numberphile: How to Trisect an Angle with Origami*. 12 December 2014.

Fibonacci, Leonardo Pisano (1202). *Liber Abaci*. English translation by Laurence Sigler (2010). *Fibonacci's Liber Abaci: A Translation into Modern English of Leonardo Pisano's Book of Calculation*.

Kaplan, R. (2000). *The Nothing that Is: A Natural History of Zero*. Oxford University Press.

Katz, Victor J. (1998), *A History of Mathematics: An Introduction* (2nd ed.), Pearson.

Pumfrey, L. (2022). *The History of Fractions*. History of Fractions (maths.org)

Seife, C. (2000). *Zero: The Biography of a Dangerous Idea*. Penguin.

Smorynski, Craig (2007), *History of Mathematics: A Supplement*, Springer, P.130.

2

The Square Root of -1

Wherein numbers become complex and the power of mathematical analysis is unleashed as a result.

In chapter 1 we saw that, as an alternative to the geometrical motivation, the requirement for irrational numbers could be motivated by insisting that an equation such as $x^2 - 2 = 0$ has a solution.

Similarly, whilst negative numbers have a natural interpretation in terms of debt, we can also motivate their creation by insisting that equations such as $x + 1 = 0$ have a solution. Alternatively, we may want to ensure that expressions like $x - y$ always have a numerical evaluation, even when y is greater than x.

This insistence that equations must have a solution, even when they do not, is a remarkably fecund mindset. The trick, when no solutions exist, is to extend the universe of possible quantities to include new entities specially tailored to provide the desired solutions. Indeed, the theme of this book is exactly that – specifically five such examples conjured by impossible square roots.

In that spirit, what is the solution to $x^2 + 1 = 0$?

Just as we insisted that $\sqrt{2}$ and -1 must exist, we now insist that $\sqrt{-1}$ must exist. We give it a symbol,

$$i = \sqrt{-1}$$

We can now solve any equation of the form $x^2 + a = 0$ where a is a positive number by assuming that this object, i, behaves like other algebraic quantities (a vague statement made precise below). This suggests that the solution to $x^2 + a = 0$ can be found as follows,

$$x^2 = -a = a \times -1 \Rightarrow x = \sqrt{a \times -1} = \sqrt{a} \times \sqrt{-1} = (\sqrt{a})i$$

We have created imaginary numbers. Any number of the form ki where k is a real number is called imaginary.

2.1 Complex Numbers

Let us continue our habit of insisting that previous arithmetical operations continue to make sense even after increasing the range of things considered as "numbers". So, we insist that $x + y$ must exist, even when our new imaginary numbers are included. If both x and y are imaginary this is easy enough, for example $3i + 5i = 8i$. But what if x is real and y is imaginary? Then there would appear to be nothing we can do to "compute" an expression like $2 + 5i$. This is a number which irreducibly has two parts, a real part and an imaginary part. Numbers of this new kind are called complex numbers.

The word "complex" does not mean "complicated". "Complex" means "consisting of more than one part". That's why complex numbers are so called – because they consist, irreducibly, of two parts. It is not because they are complicated.

The most general complex number is of the form $x + iy$ where x and y can be any real numbers.

Continuing our habit of insistence, if these new "numbers" are to qualify as a type of number then the operations of addition, subtraction, multiplication, division, taking roots, etc., must all be defined for them. This is done by requiring that all algebraic properties are shared with real numbers. This vague statement is made precise by the requirement that the complex numbers form a field, the requirements for which are specified in Appendix F. The unit element under addition, denoted 0, is defined such that $z = z + 0$ for all complex numbers, z. That means that both the real and imaginary parts of 0 are zero, and hence that it is unique.

I am not going to develop all the properties of complex numbers axiomatically, though the reader can readily check the defining requirements for a field, stated in Appendix F, are obeyed. It suffices to demonstrate a few basic properties. The definition of a field requires commutativity, so that $x + iy$ is defined as the same as $iy + x$, so we have enough information to compute the sum of two complex numbers since,

$$x + iy + a + ib = x + a + iy + ib = (x + a) + (y + b)i$$

(where the last step actually assumes that multiplication is distributive over addition, another field property). So, real and imaginary parts are simply added independently.

The notation $z = x + iy$ immediately suggests a point in a plane defined by Cartesian coordinates x and y. This complex plane is referred to as the

Argand plane. Two complex numbers represent the same point in the Argand plane iff the complex numbers are themselves equal.

Similarly, having already alluded to the definition of a field requiring that the distributive law, familiar from the algebra of real numbers, also holds, that is for any three complex numbers, a, b, c, we have $a(b + c) = ab + ac$, then we have enough information to derive the product of two complex numbers because,

$$(x + iy)(a + ib) = x(a + ib) + iy(a + ib) = ax + ibx + iay + i^2 yb$$
$$= ax - by + i(bx + ay)$$

Division and taking roots can also be done, of course, but their expression is rather messy with complex numbers in this form. They are greatly simplified when complex numbers are expressed in polar form. Let θ be the angle with respect to (henceforth, wrt) the x-axis of the point $z = x + iy$. If the 'distance' of the point from the origin is $r = \sqrt{x^2 + y^2}$ then clearly $x = r \cos \theta$ and $y = r \sin \theta$ so that,

$$z = x + iy = r \cos \theta + ir \sin \theta = re^{i\theta} \qquad (2.1.1)$$

The last step follows from the infinite series representations of the functions (which we will look at in more detail in §2.13 and §2.14 wherein the exponential function $e^{i\theta}$ and the trigonometric functions are defined and their basic properties derived). $r = \sqrt{x^2 + y^2}$ is called the "modulus", or sometimes the absolute value or magnitude of z, whilst θ is called its "argument" or sometimes its "phase".

(2.1.1) is periodic with period 2π, so the same z results if we add $2\pi n$ to θ, for any positive or negative integer n. In order to make the correspondence between x, y coordinates and r, θ coordinates single-valued we make θ unique by restricting it to be greater than or equal to zero and less than 2π. Hence $x, y \in [-\infty, +\infty]$ maps uniquely into $r \in [0, \infty], \theta \in [0, 2\pi)$. The latter angular range may be chosen to be any 2π range, and $\theta \in [-\pi, \pi)$ is perhaps the more common choice.

We shall see later that, whilst adding $2\pi n$ to θ leaves the point on the Argand plane unchanged, it changes the point on the closely related Riemann surface, a concept crucial in making sense of multi-valued functions of z (see §2.17).

A square root of $z = re^{i\theta}$ is then $\sqrt{r}e^{i\theta/2}$ (Note that I wrote "a square root", not "the square root").

In the polar expression of complex numbers the division of $z_1 = r_1 e^{i\theta_1}$ by $z_2 = r_2 e^{i\theta_2}$ is simple, it is $\frac{r_1}{r_2} e^{i(\theta_1 - \theta_2)}$.

Finally, we note that two complex numbers are equal iff their reals parts are equal and their imaginary parts are also equal. So $x + iy = a + ib$ means that both $x = a$ and $y = b$. Alternatively, $r_1 e^{i\theta_1} = r_2 e^{i\theta_2}$ iff $r_1 = r_2$ and $\theta_1 = \theta_2$ modulo 2π.

2.2 Quadratic Equations and Complex Numbers

Now here is another topic which the unmathematical regard as designed only to torture schoolchildren. You will have learnt 'by rote' that the *two* solutions to the quadratic equation,

$$az^2 + bz + c = 0 \qquad (2.2.1)$$

for non-zero a, are,

$$z = \frac{-b \pm \sqrt{b^2 - 4ac}}{2a} \qquad (2.2.2)$$

You may check that this is correct by substitution into the quadratic, (2.2.1), and showing that the result is identically zero, whichever sign is chosen. Our insistence that a quadratic always has a solution, no matter what values are assigned to the coefficients a, b, c leads immediately to the need for complex numbers, because (2.2.2) is complex if the so-called discriminant, $b^2 - 4ac$ is negative.

But we can also turn this around. How would one derive the solution (2.2.2), say if you had forgotten it and had no source reference to hand? For this purpose we initially assume a, b, c are real, but that the solution $z = x + iy$ is complex. Substitution of $z = x + iy$ into (2.2.1) gives,

$$a(x^2 - y^2 + 2ixy) + b(x + iy) + c = 0$$

But, if a complex quantity is zero, both its real and imaginary parts must be zero, so this splits into two equations,

$$a(x^2 - y^2) + bx + c = 0 \text{ and } 2axy + by = 0$$

The second of these implies that either $y = 0$ or $x = -b/2a$. Assuming the latter and substituting into the first equation gives,

$$ay^2 = a\left(\frac{-b}{2a}\right)^2 + b\left(\frac{-b}{2a}\right) + c$$

Hence,
$$y^2 = \frac{-b^2+4ac}{4a^2} \text{ or } y = \pm\sqrt{\frac{-b^2+4ac}{4a^2}} = \pm\frac{\sqrt{-b^2+4ac}}{2a}$$

Hence, the solution is, as required by (2.2.2),

$$z = x + iy = \frac{-b}{2a} \pm \frac{i\sqrt{-b^2 + 4ac}}{2a} = \frac{-b \pm \sqrt{b^2 - 4ac}}{2a}$$

Note that we have absorbed the factor of i by using its equivalence to $\sqrt{-1}$. The cunning of this trick is that, despite the logic of the derivation being based upon the assumption that the complex part (y) is non-zero, the resulting expression is still correct even when it is zero! In fact, the logic of the derivation also requires that the coefficients a, b and c are real, but the solution, (2.2.2), applies also when these are themselves complex – as is demonstrated simply by substitution.

A type of generalisation of this trickery works also for the cubic equation (see Appendix G), and perhaps for the quartic, though I've never had the strength to attempt it. However it certainly fails for the general quintic and above. Historically, mathematicians sought a corresponding solution for the quintic "in surds" (sometimes called "radicals", i.e., expressions raised to a rational power, including square-roots, cube-roots, etc.). They sought in vain for many years before realising that no such expression exists. The proof that the quintic could not be solved in surds preceded the understanding of why this was so. Abel was the first to publish a complete proof in 1824 (following a flawed proof by Ruffini in 1799). Some sources claim that Galois had found a proof before Abel. This is false. Galois was only 12 in 1824.

The underlying reason why general polynomials of order 5 and above have no solution in surds (or radicals) was, however, clarified by Galois theory, which is now the standard way of approaching these matters. The modern theory of rings, fields and groups grew from Galois theory. A hopelessly incomplete, but relatively painless, sketch of fields and field extensions, which are a precursor to Galois Theory, is provided in Appendix F. Such material is covered in undergraduate courses in pure mathematics, but as an ignorant peasant physicist, I draw a veil over my ignorance at this point.

2.3 Polynomial Equations and Algebraic Closure

The insistence that certain equations have solutions (also, of course, called "roots") has proved so effective in extending what is meant by "a number" that the obvious next step is to ask whether we get any further extensions if we want to ensure that every polynomial equation has at least one root. In other words, consider the following equation, and assume it does not reduce simply to a constant (i.e., that at least one of the other terms is non-zero),

$$A_n z^n + A_{n-1} z^{n-1} + \cdots A_1 z + A_0 = 0 \qquad (2.3.1)$$

If the coefficients, A_r, are arbitrary complex numbers, does (2.3.1) always have at least one complex root, i.e., is there at least one complex number z for which (2.3.1) holds for arbitrary complex A_r?

The answer is "yes", and this is actually a definition of what is meant by the "algebraic closure" of complex numbers. The complex numbers are said to form an "algebraically closed field". Thus our programme of generating numbers of ever more general type, through solutions to equations of polynomial type, comes to an end at this point. We have achieved closure.

The real numbers are not algebraically closed. A polynomial equation with real coefficients does not always have a real root, e.g., a quadratic with a negative discriminant.

Similarly, rational numbers are not algebraically closed, either – a statement which is essentially a paraphrasing of chapter 1.

On the other hand, the algebraic numbers *are* algebraically closed. This is not obvious. Algebraic numbers are defined as the roots of polynomial equations with rational coefficients. It is not obvious that any polynomial equation with algebraic coefficients has an algebraic root – but it is so. In fact, the algebraic numbers are the smallest algebraically closed field which contains the rational numbers. The algebraic number field is therefore said to be the algebraic closure of the rationals.

2.4 The Fundamental Theorem of Algebra

The Fundamental Theorem of Algebra is so-called by analogy with the Fundamental Theorem of Arithmetic. Recall that the latter relates to the unique factorisation of an integer into prime factors. One of the many different ways of expressing the Fundamental Theorem of Algebra is that any polynomial with complex coefficients,

$$A_n z^n + A_{n-1} z^{n-1} + \cdots A_1 z + A_0$$

with non-zero A_n, is identical to a factorised form,

$$A_n (z - z_n)(z - z_{n-1}) \dots (z - z_1) \qquad (2.4.1)$$

in which the n quantities z_r are complex numbers (noting that real numbers are a special case of complex numbers). It is immediately obvious that the polynomial equation obtained by setting the polynomial to zero therefore has precisely n roots (solutions), namely $z = z_r$ for $r = 1,2,3$ .... (Noting that some of these roots may be equal).

The naming of the Fundamental Theorem of Algebra (henceforth FTA) is rather an historical misnomer. It is understandable how this arose by analogy with the Fundamental Theorem of Arithmetic. However, as regards modern algebra, it is not really fundamental. And, whilst it is an algebraic statement, it is not a theorem which is algebraically provable. Statements, and purported proofs of the theorem (often flawed) span the last 400 years up to the modern era.

Correctly understood, the FTA is actually a restatement of the closure of the complex number field. The reason is that, if we know – from algebraic closure – that there is at least one root of (2.3.1), say z_n, then there is an effective procedure to find a polynomial $Q(z)$ of order $n - 1$ such that,

$$A_n z^n + A_{n-1} z^{n-1} + \cdots A_1 z + A_0 \equiv (z - z_n) Q(z) \qquad (2.4.2)$$

But $Q(z) = 0$ must also have at least one complex root, by the same closure theorem, say z_{n-1}, and so the original polynomial now becomes $(z - z_n)(z - z_{n-1})R(z)$ for some polynomial $R(z)$ of order $n - 2$, and so on until the whole polynomial is "unzipped" into the fully factorised form of (2.4.1).

Consequently, the proof of the FTA consists of proving two things: firstly, the closure of the complex field, i.e., that there is at least one root of every polynomial equation with complex coefficients, and also demonstrating the existence of the effective procedure which yields $Q(z)$ in (2.4.2).

I will give a proof of closure in §2.12. It is a rather beautiful application of complex function theory, specifically Cauchy's integral theorem. Alternative proofs of the closure of the complex field abound, several different proofs based on analysis, and also proofs based on induction, on topology, on geometry, on linear algebra or using Galois theory.

Here I will show how to construct the quotient polynomial $Q(z)$ in (2.4.2). This has the advantage of also illustrating why a purely algebraic proof of FTA is not possible.

Without loss of generality we can take $A_n = 1$ (i.e., we can divide through the polynomial by A_n as it is non-zero by assumption). Consequently we want to find the polynomial,

$$Q(z) = z^{n-1} + a_{n-2}z^{n-2} + a_{n-3}z^{n-3} + \cdots + a_1 z + a_0$$

such that, for some given b,

$$z^n + A_{n-1}z^{n-1} + \cdots A_1 z + A_0 \equiv (z - b)Q(z) \qquad (2.4.3)$$

By equating coefficients of powers of z the relationship between the two sets of coefficients is seen to be,

$$A_{n-1} = -b + a_{n-2}$$
$$A_{n-2} = -ba_{n-2} + a_{n-3}$$
$$A_{n-3} = -ba_{n-3} + a_{n-4}$$

to,

$$A_1 = -ba_1 + a_0$$
$$A_0 = -ba_0$$

To find the quotient polynomial, $Q(z)$, i.e., the coefficients a_r, given the coefficients A_r and the root b, the first of the above equations provides a_{n-2}. Substituting that into the second equation then provides a_{n-3}, and so on, until a_0 is determined by the second-to-last equation. The last equation is "surplus to requirements", but is guaranteed to also be obeyed if the given b is indeed a valid root as assumed. Hence, if we know – from the FTA – that there is a root b then the above procedure guarantees to find the quotient polynomial, $Q(z)$, and, by repetition, the fully factorised form of (2.4.1) results.

However, if we attempt to prove the FTA using this algebraic procedure, it fails. In that case we would not know that any b exists such that (2.4.3) holds. The simple solution procedure fails because both the first and the last equations relating the coefficients involves two unknowns. It is easy to successively eliminate the coefficients a_r to leave an equation in only b, but this is just the original polynomial, $b^n + A_{n-1}b^{n-1} + \cdots A_1 b + A_0 = 0$, which merely tells us that if we assume that there is some b which is a root, then….it's a root! So we get nowhere.

My analysis-based proof of the FTA will follow shortly, after setting up some key analytical results relating to differentiation and integration. It is remarkable that these considerations, apparently far removed from polynomials and algebraic closure, permit the proof to be performed in an elegant fashion. It illustrates that mathematics is not only unreasonably effective in physics, but mathematics is unreasonably effective in mathematics too.

2.5 Mathematical Analysis

I may have annoyed some readers in preceding sections by using the term "analysis", meaning a branch of mathematics, without defining it. From the context you will note my drawing a clear distinction between analysis and other branches of mathematics, such as geometry, topology, algebra, number theory and logic. I won't attempt a really rigorous definition of analysis, if indeed there is one. It suffices to say that analysis deals with functions, both real and complex, their differentiation and integration (i.e., calculus), and the properties of equations and solutions to equations, especially differential equations. Infinite series are part of analysis because some functions (namely the transcendental functions) are defined by infinite series. Consequently, the issue of convergence of infinite series is also part of analysis.

For physicists (and engineers) it is analysis which is the most relevant branch of mathematics. Indeed physicists often have only the vaguest notion of the branches of mathematics other than algebra and analysis, and a smattering of geometry. (I say this as a physicist). For IT specialists, it will be otherwise, with finite mathematics and logic being of primary significance.

Calculus and complex functions are the foundations of analysis, without which it would hardly exist. I will be assuming a knowledge of calculus.

2.6 Differentiation of Complex Functions

By a "complex function" we mean a complex valued function of a complex variable, $f(z)$. Like real functions, we could also consider complex functions of several variables, but we shall be content here with functions of just one complex variable. The derivative is defined as for real functions,

$$\frac{df}{dz} = \lim_{z \to z_0} \frac{f(z) - f(z_0)}{z - z_0} \tag{2.6.1}$$

We cannot assume for an arbitrary function that the derivative exists. Moreover, there is a problem with the definition (2.6.1). The same problem arises in less severe form with real functions. What if the graph of a real function has a "kink" at z_0 so that its gradient is discontinuous at that point? Then the derivative is not unique at that point, as the limiting process in (2.6.1) gives a different result according to whether z_0 is approached from above or below. Often, in applications, this is addressed by excluding by *fiat* any such functions, that is, by assuming they are differentiable (at least once, and perhaps many times). However, this will not be an option if the function is defined by other means, and so some functions are just not differentiable at some points.

For complex functions the requirement for unique differentiability is more demanding because the point z_0 can now be approached, not just from two directions, but from any direction in the Argand plane. To define a unique quantity, the limiting process in (2.6.1), i.e., $z \to z_0$, must provide the same result for the derivative regardless of the direction in the Argand plane from which z approaches z_0. This is a very significant restriction on the type of function for which the derivative, in this sense, exists.

A complex function for which the derivative exists at point z is said to be holomorphic at the point. A function may be said to be holomorphic over a certain region Ω of the Argand plane if it is holomorphic at every point (i.e., for every complex number) in that region. A function which is holomorphic everywhere in the Argand plane is called an entire function. Examples are polynomials or the exponential function (which therefore includes sines, cosines and their hyperbolic counterparts, see §2.14).

It is often the case that functions are holomorphic almost everywhere except at isolated points – but that these isolated points have great significance for the function overall.

Some important results follow immediately from the uniqueness of the derivative. Suppose we write $f = u + iv$, where u and v are the real and imaginary parts of f respectively (and both depend upon both x and y, the real and imaginary parts of z). We can consider z approaching z_0 parallel to the x-axis, so $z - z_0 = \delta x$, or parallel to the y-axis so $z - z_0 = i\delta y$. The resulting derivative must be the same in these two cases, so that,

$$\frac{\delta_x u + i\delta_x v}{\delta x} = \frac{\delta_y u + i\delta_y v}{i\delta y}$$

i.e.,

$$\frac{\partial u}{\partial x} + i\frac{\partial v}{\partial x} = \frac{1}{i}\cdot\frac{\partial u}{\partial y} + \frac{\partial v}{\partial y}$$

Equating real and imaginary parts, and recalling that $\frac{1}{i} = -i$ we have,

$$\frac{\partial u}{\partial x} = \frac{\partial v}{\partial y} \quad \text{and} \quad \frac{\partial v}{\partial x} = -\frac{\partial u}{\partial y} \tag{2.6.2}$$

Hence, the x and y derivatives of the real and imaginary parts are related in a non-trivial (indeed, highly restrictive) manner. The conditions (2.6.2) are known as the Cauchy-Riemann equations. We have shown that these Cauchy-Riemann equations necessarily follow from a function being holomorphic. Given some minimal continuity assumptions, the Cauchy-Riemann equations are also sufficient to ensure the function is holomorphic.

Taking the x derivative of one of (2.6.2) and the y derivative of the other and adding or subtracting shows that both real and imaginary parts obey the 2-dimensional Laplace equation,

$$\left(\frac{\partial^2}{\partial x^2} + \frac{\partial^2}{\partial y^2}\right)u = 0 \quad \text{and} \quad \left(\frac{\partial^2}{\partial x^2} + \frac{\partial^2}{\partial y^2}\right)v = 0 \tag{2.6.3}$$

So, a holomorphic function automatically obeys Laplace's equation (in 2D). But the most stunning result of being holomorphic, i.e., having a unique derivative, is Cauchy's integral theorem to which we now turn.

2.7 The Holomorphic Integral Theorem

The Holomorphic Integral Theorem states that the integral of any function $f(z)$ around any closed path in the Argand plane is zero if the function is holomorphic everywhere on and within the integration path (assuming the enclosed region is simply connected, i.e., it has no holes). The integral is with respect, of course, to dz where $z = x + iy$, so that $dz = dx + idy$. Hence the theorem is,

$$\oint_{\delta C} f(z)dz = 0 \tag{2.7.1}$$

where δC denotes the integration path, which must be closed. The theorem is easily proved by making use of Green's Theorem which, for two functions g and h of x, y, relates a line integral around a closed path δC to an integral over the area enclosed by this path, C. Specifically,

$$\oint_{\delta C}(g\,dx + h\,dy) = \iint_C \left(\frac{\partial h}{\partial x} - \frac{\partial g}{\partial y}\right)dx\,dy \qquad (2.7.2)$$

For completeness this theorem is proved in Appendix H. The integral (2.7.1) may be written in terms of real and imaginary parts of $f = u + iv$ as,

$$\oint_{\delta C} f(z)\,dz = \oint_{\delta C} (u + iv)(dx + idy)$$

$$= \oint_{\delta C} [(u\,dx - v\,dy) + i(v\,dx + u\,dy)]$$

Green's Theorem allows the real part to be written,

$$\oint_{\delta C} (u\,dx - v\,dy) = \iint_C \left(-\frac{\partial v}{\partial x} - \frac{\partial u}{\partial y}\right)dx\,dy$$

And the imaginary part becomes,

$$\oint_{\delta C} (v\,dx + u\,dy) = \iint_C \left(\frac{\partial u}{\partial x} - \frac{\partial v}{\partial y}\right)dx\,dy$$

But the Cauchy-Riemann equations, (2.6.2), which apply when the function is holomorphic, show that both these integrands are zero everywhere within the region C, and on its boundary, provided that the function is holomorphic throughout this region. Hence the integral is always zero for such a holomorphic function. QED.

2.8 Cauchy's Integral Theorem

What if we deliberately make the integrand non-holomorphic at a single point by dividing the otherwise holomorphic function by $(z - z_0)$ for some complex constant, z_0? What can we say then about the integral,

$$\oint_{\delta C} \frac{f(z)}{z - z_0}\,dz \qquad (2.8.1)$$

Well, the integrand is, in fact, holomorphic everywhere on and within δC if the point z_0 lies outside this region. Hence,

For z_0 outside δC: $\qquad \oint_{\delta C} \frac{f(z)}{z - z_0}\,dz = 0 \qquad (2.8.2)$

We cannot conclude this, however, if z_0 lies inside δC. However, even in this case we can conclude that the integration contour, δC, can be changed to any other integration contour, $\delta C'$, without changing the integral so long as both contours enclose the point z_0. This is because the function is holomorphic in the region between the two contours. Consider Figure 2.1 in which both contours δC and $\delta C'$ are counter-clockwise. The combined contour defined by $\delta C + A - \delta C' + B$ encloses a region within which the integrand $\frac{f(z)}{z-z_0}$ is holomorphic (because the point z_0 has been excluded). Note that the contour $\delta C'$ contributes with a minus sign here because it is necessary to traverse it clockwise to make the combined contour continuous. In the limit that the lines joining the two curved contours overlap, $A \to B$, their contribution cancels (because the two integrals are conducted in opposite directions), and in this limit the contours δC and $\delta C'$ become closed. Hence, in this limit the integral of $\frac{f(z)}{z-z_0}$ over $\delta C - \delta C'$ is zero, i.e., its integral over δC equals that over $\delta C'$.

Hence, when z_0 lies inside δC the integral (2.8.1) is independent of the contour used provided that z_0 lies inside whatever contour is used. Consequently we may shrink the contour to a tiny circle centred on point z_0. If we shift the origin to the point z_0 then $z - z_0 \to z = re^{i\theta}$ whilst $dz = d\left(re^{i\theta}\right) = ire^{i\theta}d\theta$ because we are considering now a circular contour on which r is constant. Hence the Cauchy integral becomes,

$$\oint_{\delta C} \frac{f(z)}{re^{i\theta}} ire^{i\theta} d\theta = i \oint_{\delta C} f(z)\, d\theta$$

Figure 2.1: Contour independence of the Cauchy integral

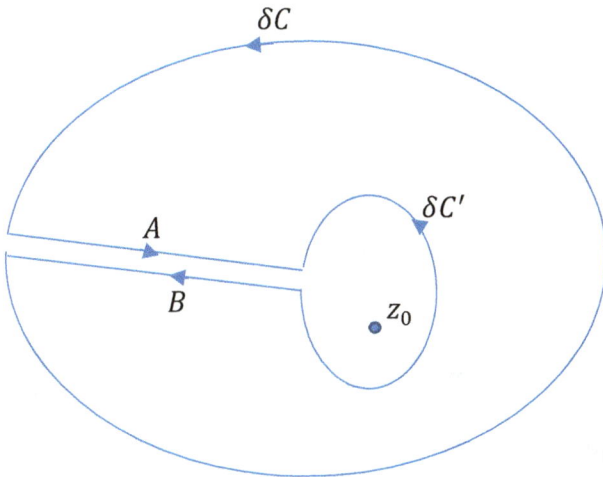

But as we let the circular contour shrink to an arbitrarily small radius, centred on z_0, the function becomes the constant $f(z_0)$. The integral is just the total angle of a complete circle, i.e., 2π. Hence, finally, we get the Cauchy Integral Formula,

For z_0 inside δC:
$$\oint_{\delta C} \frac{f(z)}{z-z_0}\, dz = 2\pi i f(z_0) \tag{2.8.3}$$

The remarkable thing about this result is that the integral is evaluated using only the values of the integrand, and hence only the values of f, on the boundary of the region in question – and this may be at an arbitrarily large distance from the point z_0. But the result is simply a constant $(2\pi i)$ times the value of the function at z_0.

2.9 Holomorphic Functions are Analytic Functions

I confess that for many years I used the terms "holomorphic" and "analytic" interchangeably. Fortunately, in the context of complex functions, they are equivalent. However, whereas a holomorphic function is defined by the existence of the first complex derivative, an analytic function is defined by the existence of a representation of the function within some region by a convergent power series, i.e.,

$$f(z) = \sum_{r=0}^{\infty} a_r (z - z_0)^r \tag{2.9.1}$$

It follows that an analytic function is infinitely differentiable because the above series can be identified with the Taylor series and the coefficients given by,

$$a_r = \frac{1}{r!} \frac{\partial^r f}{\partial z^r}\bigg|_{z_0} = \frac{f^{(r)}(z_0)}{r!} \tag{2.9.2}$$

That a complex analytic function is holomorphic is clear, i.e., the first derivative exists because all the derivatives exist. However, what is far from obvious is that the reverse is also true: a holomorphic function is a complex analytic function. The existence of the first derivative is sufficient to imply the existence of all derivatives. (This is not true for real functions, of course, and is an example of how much more restrictive is the existence of the derivative in the complex case).

The analyticity of holomorphic functions may be established using Cauchy's Integral Formula, which can be written, for some dummy complex integration variable, w,

$$f(z) = \frac{1}{2\pi i} \oint \frac{f(w)}{w-z}\, dw$$

The contour of integration has not been made explicit but must contain the point z. Note that we can write,

$$\frac{1}{w-z} = \frac{1}{w-z_0} \cdot \frac{1}{\left(1 - \frac{z-z_0}{w-z_0}\right)} = \frac{1}{w-z_0} \sum_{r=0}^{\infty} \left(\frac{z-z_0}{w-z_0}\right)^r$$

The last step is valid only if $|z - z_0| < |w - z_0|$ so the common ratio of the geometric progression is less than one, and hence that the sum converges. This is guaranteed to be the case if we restrict z_0 to be arbitrarily close to z, which will be sufficient for the proof. Hence,

$$f(z) = \frac{1}{2\pi i} \sum_{r=0}^{\infty} (z - z_0)^r \oint \frac{f(w)}{(w-z_0)^{r+1}} dw \qquad (2.9.3)$$

Real mathematicians will question the veracity of interchanging the order of the sum and the integral – I'll leave it as an exercise for the reader to show that this is valid – it is. This is job done as (2.9.3) is in the form of (2.9.1) with the expansion coefficients given explicitly by,

$$a_r = \frac{1}{2\pi i} \oint \frac{f(w)}{(w-z_0)^{r+1}} dw \qquad (2.9.4)$$

The restriction noted above that z_0 must be arbitrarily close to z for this derivation to be valid means that we have established only that the power series, (2.9.1), exists (i.e., is convergent) in a very small neighbourhood of z. However, the argument applies at any point z at which the function is holomorphic. Hence, if the function is holomorphic throughout an extended domain then the power series exists everywhere within that domain too.

2.10 Cauchy's Integral Formula for Derivatives

This follows immediately from the previous section as (2.9.2) and (2.9.4) are both expressions for the same expansion coefficients so we conclude that the rth derivative at point z_0 can be written as a path integral analogous to the Cauchy Integral Formula for $f(z_0)$,

$$f^{(r)}(z_0) = \frac{r!}{2\pi i} \oint \frac{f(w)}{(w-z_0)^{r+1}} dw \qquad (2.10.1)$$

assuming z_0 is inside the integration contour. Just as Cauchy's Integral Formula picks out the value of the function at the zero of the denominator, the modified form of (2.10.1), with a higher power of the denominator, picks out the rth derivative at the zero of the denominator.

2.11 Liouville's Theorem

Perhaps the most unexpected theorem in complex analysis, in a sequence of unexpected theorems, is Liouville's Theorem. It states that any entire function which is bounded everywhere in the Argand plane is necessarily a constant. Recall that an entire function is a function which is holomorphic everywhere in the Argand plane. "Bounded" means, of course, that there is some finite number B such that $|f(z)| < B$ at all points z across the whole Argand plane. It's hard to believe at first. The corollary of Liouville's Theorem is that any function which is holomorphic everywhere but not a constant must blow-up somewhere (often at infinity).

The proof is simple and proceeds from our knowledge of the expansion coefficients in (2.9.1), which must hold everywhere for entire functions, namely (2.9.4). Because we are assuming $|f(z)| < B$ everywhere we note that,

$$|a_n| = \left| \frac{1}{2\pi i} \oint \frac{f(w)dw}{(w-z_0)^{n+1}} \right| \leq \frac{1}{2\pi} \oint \frac{|f(w)||dw|}{|w-z_0|^{n+1}} \leq \frac{1}{2\pi} \oint \frac{B|dw|}{r^{n+1}} = \frac{2\pi r B}{2\pi r^{n+1}} = \frac{B}{r^n}$$

where we have exploited the contour independence of the integral and have chosen to evaluate on a circle of radius r centred on the point z_0 about which the expansion (2.9.1) was defined. Since the function is holomorphic everywhere, by assumption, the above conclusion continues to hold if we let $r \to \infty$ in which case we see that all the coefficients a_n are zero except for a_0, i.e., the function must be a constant. QED.

Another corollary of Liouville's Theorem is that, for some z, "almost all" points on the Argand plane can be obtained as an entire function $f(z)$. [To stop mathematicians squealing too much, to be precise I should say "the image of f is dense in the complex field"]. To prove this by *Reductio ad Absurdum*, suppose there was a complex number, w, such that whatever z we choose we could never have $f(z) = w$, and, in fact, the closest we could get maintained a finite, non-zero 'distance' between $f(z)$ and w, i.e., for all z, there is some real number A such that $|f(z) - w| > A$. Consider the function,

$$g(z) = \frac{1}{(f(z)-w)}$$

Since the denominator cannot be zero, by assumption, and because $f(z)$ is assumed entire, it follows that the derivative of $g(z)$ always exists, everywhere, and hence g is entire. But because its denominator is bounded below, $|g|$ is bounded above. So g is a bounded entire function and hence must be a constant, which means f must be a constant, which is a contradiction. Hence, contrary to our assumption, it must be that for any w we can always find some z such that, for any non-constant, entire function, $f(z)$ is arbitrarily close to w.

Thus, non-constant, entire functions take values which cover "the whole" Argand plane (strictly "are dense within the Argand plane"). Note that this subsumes such functions blowing up somewhere, since they are obliged to reproduce the points at infinity on the Argand plane.

2.12 Proof of the Fundamental Theorem of Algebra

Finally, a proof of the Fundamental Theorem of Algebra (FTA). We seem to have come a long way since considering the apparently elementary issue of factoring polynomials – through some seemingly heavy-duty theorems about differentiation and integration – though, actually, all these theorems were very easy to prove. Why would such sophisticated theorems in analysis even be relevant to the (apparently) purely algebraic issue of factoring polynomials?

As hinted previously, it is because, whilst the factoring of polynomials is algebraic in appearance, its veracity depends upon the "completeness", in some sense, of the number field in question. This is clear because polynomials cannot always be factored as in (2.4.1) if we confine ourselves to real numbers, even if the coefficients are real. Proof of the FTA must, therefore (assuming it is true) draw upon properties of complex numbers not shared by real numbers. Consequently, neither the theorem nor the proof of it can be purely algebraic because the reals and the complex numbers share the same algebraic properties.

In contrast, it turns out that the specific properties of complex numbers which are distinct from those of the reals are sufficiently codified within the theorems about holomorphic functions (derived above) to put us in a position to prove the FTA.

Recall that we showed in §2.4 that to establish the FTA we need only demonstrate closure, i.e., that there is always one complex root of any polynomial with complex coefficients. We can show this very easily using Liouville's Theorem. Suppose $f(z)$ is a nth order polynomial. Without loss of generality we can assume the leading coefficient is unity for $n \geq 1$. Hence,

$$f(z) = z^n + A_{n-1}z^{n-1} + \cdots A_1 z + A_0$$

Now define,

$$g(z) = \frac{1}{f(z)}$$

Proceed once again by *Reductio ad Absurdum* by assuming $f(z)$ has no zero for any complex number. In this case $g(z)$ is an entire function because its derivative exists everywhere, namely $g' = -f'/f^2$. That $f(z)$ has no zero is key to this derivative existing everywhere. But $g(z)$ is also bounded, because if it were unbounded somewhere (i.e., $g \to \infty$ at some point) then f would be zero at that point, contrary to assumption. Hence, g is a bounded entire function and hence, by Liouville's Theorem, g is constant. But that means f is constant – which is a contradiction. Hence the assumption that $f(z)$ has no zero must be false and we conclude that a non-constant polynomial must always have a zero. From §2.4 this is all we need to conclude that the FTA is true and polynomials always factor as $(z - z_n)(z - z_{n-1}) \ldots (z - z_1)$ for some complex numbers (roots) z_r.

The immediate corollary is that an nth order polynomial always has exactly n roots, in general complex, although some of these roots might be equal. Any claim to have found $n + 1$ different roots of an nth order polynomial must be false.

A further simple corollary is that polynomials with real coefficients have complex roots which occur in conjugate pairs. This follows because taking the conjugate of $z^n + A_{n-1}z^{n-1} + \cdots A_1 z + A_0 = 0$ gives exactly the same equation but with z replaced by z^*. So, if $a + ib$ is a root, with real a, b, then $a - ib$ is a root of the same polynomial.

Yet another immediate corollary is that if the order of a polynomial with real coefficients is odd, then it must have at least one real root.

2.13 The Exponential Function

So, finally, we get to define the exponential function. In the usual school pedagogy one is exposed to the trigonometric functions, logarithms and the exponential function at a tender age. But only people who go on to read mathematics or physics at university level will ever meet the theory of holomorphic functions and Cauchy's Integral Theorem. Yet in my development I have radically reversed the order. There is method in this madness as only when the more sophisticated structures are in place can these familiar transcendental functions be appreciated in their proper context.

The exponential function is an example of an entire function which is not a polynomial. It is also an extremely important function, none more so in mathematical physics. It is a transcendental function, i.e., it is an analytic function which cannot be expressed in terms of a finite sequence of the algebraic operations of addition, subtraction, multiplication, division, raising to a power, and root extraction. Transcendental functions are generally defined by an infinite series. The exponential function can be defined in this way, but it is more elegantly defined as the function which equals its own derivative,

$$\frac{df}{dx} = f \tag{2.13.1}$$

and for which $f(0) = 1$. Assuming the function defined this way has a Taylor series,

$$f(x) = \sum_{n=0}^{\infty} a_n x^n \tag{2.13.2}$$

According to (2.13.1) this must also be identical to $\sum_{n=1}^{\infty} n a_n x^{n-1}$ from which the reader should check we get the recursion relation $a_{n+1} = a_n/(n+1)$, which therefore gives $a_{n+1} = 1/n!$. So the power series for the exponential function, which I now write in terms of a complex variable, z, is,

$$f(z) = \sum_{n=0}^{\infty} \frac{z^n}{n!} = 1 + z + \frac{z^2}{2} + \frac{z^3}{3!} + \frac{z^4}{4!} + \dots \tag{2.13.3}$$

I leave it as an exercise for the reader to show that this series is absolutely convergent, i.e., it converges for all finite complex z.

Napier's constant e (also called Euler's number, though I think Euler has enough credits to his name) is defined as $f(1)$. Hence,

$$e = 1 + 1 + \frac{1}{2} + \frac{1}{3!} + \frac{1}{4!} + \dots = 2.718281828459 \dots \tag{2.13.4}$$

That e is irrational, and indeed transcendental, is proved in Appendices B and D.

The reader will note that I have not yet written the exponential function in its usual form, namely as $f(z) \equiv e^z$. This is because I have not yet justified that this notation – "*e* raised to the power *z*" - is valid. Indeed, the full justification is quite problematic, an issue that will exercise us shortly. In any case, whilst (2.13.3) provides a unique $f(z)$ for any complex z, what does it mean to raise *e* to an imaginary, or a complex, power? Do not be tempted to use the fact that the derivative of e^z is e^z to justify the notation because that would be circular logic, i.e., we have not yet shown that that is the case.

By squaring (2.13.3) the reader should be able to show that the first few terms can be rearranged to be,

$$f(z)^2 = 1 + (2z) + \frac{(2z)^2}{2} + \frac{(2z)^3}{3!} + \frac{(2z)^4}{4!} + \ldots \qquad (2.13.5)$$

So the claim that $f(z) \equiv e^z$ has some credibility, because if so we expect $f(z)^2 \equiv (e^z)^2 = e^{2z} = f(2z)$ and (2.13.5), if correct to all orders, shows that is the case.

To prove this, consider the more general claim that $f(z)f(w) \equiv f(z+w)$. This is established by noting that (2.13.3) can be obtained as the limit,

$$f(z) = \lim_{n \to \infty} \left(1 + \frac{z}{n}\right)^n \qquad (2.13.6)$$

as can be checked using the binomial theorem. Hence,

$$f(z)f(w) = \lim_{n \to \infty} \left(1 + \frac{z}{n}\right)^n \left(1 + \frac{w}{n}\right)^n$$

But
$$\left(1 + \frac{z}{n}\right)^n \left(1 + \frac{w}{n}\right)^n = \left(1 + \frac{z}{n} + \frac{w}{n} + \frac{zw}{n^2}\right)^n$$

And in the limiting process the quadratic denominator of the last term makes it vanish, giving, for any complex z and w,

$$f(z)f(w) = \lim_{n \to \infty} \left(1 + \frac{z+w}{n}\right)^n = f(z+w) \qquad (2.13.7)$$

So we have $f(z)f(z) = f(2z)$, using which gives

$$f(z)f(2z) = f(z)f(z)\, f(z) = f(3z)$$

and so on, resulting in n factors of $f(z)$ giving us, $f(z)^n = f(nz)$. Setting $z = 1$ and noting that we have already identified $f(1)$ with e therefore gives us,

$$f(n) = e^n \qquad (2.13.8)$$

It is very tempting to conclude that the similar looking $f(z) = e^z$ holds for any complex z, despite (2.13.8) having been proved only for integers, n. After all, (2.13.7) is then perfectly natural, namely,

$$f(z)f(w) = e^z e^w = e^{z+w} = f(z+w)$$

So why would anyone object to writing the exponential function, defined by 2.13.3, as $f(z) \equiv e^z$? The reader's puzzlement will only be exacerbated by the simple fact that e^z is indeed the universally accepted way in which the exponential function is conventionally written. Nevertheless, it is misleading and, without caution, can lead you into all sorts of paradoxes.

Readers probably feel fully confident that they understand the innocent looking e^x or x^y when x and y (and, of course, e) are all real. These can surely hold no nasty surprises – can they? But exponentiation is more tricky than you might have thought. For example, even for real x, a, b it is treacherous to assume as an identity $(x^a)^b = x^{ab}$ without further elaboration as to what it means. Here is a counterexample,

$$((-1)^2)^{\frac{1}{2}} = 1^{\frac{1}{2}} = 1$$

whereas $(x^a)^b = x^{ab}$ implies $((-1)^2)^{\frac{1}{2}} = (-1)^{2 \times \frac{1}{2}} = (-1)^1 = -1$. The alert will spot that we could have chosen $1^{\frac{1}{2}} = -1$ to make the two consistent. But we did not have to choose that root, whereas it is indisputable that $(-1)^1$ can only be -1.

Destabilised? You should be. We return to this theme in §2.16 after a short digression.

2.14 The Trigonometric Functions

Figure 2.2: Definition and derivatives of the trigonometric functions

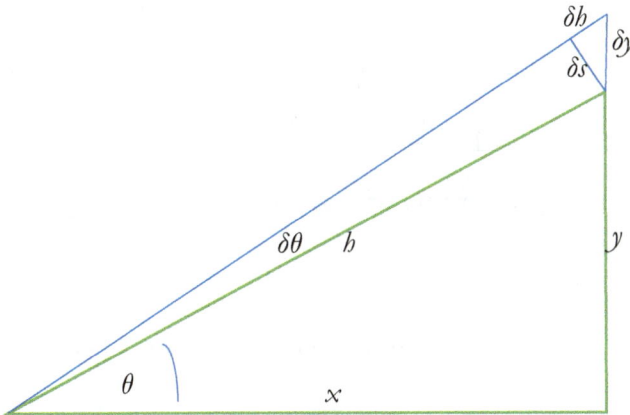

The relationship between the exponential function and the trigonometric functions, sine and cosine, can be achieved simply by defining the latter by their power series. However, this alone is not satisfactory as it fails to establish the geometric interpretation of these functions, which is generally how one first meets them. The route to linking the exponential function to geometry thus starts by defining the sine and cosine using the familiar right-angled triangle (in Euclidean space, of course),

Defining, $\qquad \sin\theta \equiv \dfrac{y}{h} \qquad$ and $\qquad \cos\theta \equiv \dfrac{x}{h}$

This immediately gives us $sin^2\,\theta + cos^2\theta \equiv 1$ by Pythagoras.

We next obtain the familiar expressions for the derivatives of these functions using the geometry (Figure 2.2). A small increment in the angle θ is considered by increasing y at fixed x. Since, by Pythagoras, $h^2 = x^2 + y^2$ and x is held constant we get $h\delta h = y\delta y$, i.e., $\delta h = \sin\theta \cdot \delta y$. Hence,

$$\delta s^2 = \delta y^2 - \delta h^2 = \delta y^2(1 - sin^2\theta) = \delta y^2 cos^2\theta \rightarrow \delta s = \delta y \cdot cos\theta$$

$$\delta\theta = \frac{\delta s}{h} = cos\theta \cdot \frac{\delta y}{h}$$

$$\delta\left(\frac{y}{h}\right) = \frac{\delta y}{h} - \frac{y}{h^2}\delta h = \frac{\delta y}{h} - \frac{y}{h^2}\sin\theta \cdot \delta y = \frac{\delta y}{h}(1 - sin^2\theta) = \frac{\delta y}{h}cos^2\theta$$

The derivative of $\sin\theta$ is thus, in the limit,

$$\frac{d\sin\theta}{d\theta} = \frac{\delta(\sin\theta)}{\delta\theta} = \frac{h}{cos\theta \cdot \delta y}\delta\left(\frac{y}{h}\right) = \frac{h}{cos\theta \cdot \delta y}\frac{\delta y}{h}cos^2\theta = cos\,\theta$$

...the familiar result. The derivative of $cos\,\theta$ can be shown to be $-sin\,\theta$ in similar manner. Armed with this we can derive the Taylor series for the trig functions, thus,

$$\sin\theta = \theta - \frac{\theta^3}{3!} + \frac{\theta^5}{5!} - \frac{\theta^7}{7!} + \cdots \qquad (2.14.1)$$

$$\cos\theta = 1 - \frac{\theta^2}{2!} + \frac{\theta^4}{4!} - \frac{\theta^6}{6!} + \cdots \qquad (2.14.2)$$

Finally, it becomes a mere matter of separating even and odd powers to see from (2.13.3) that,

$$e^{iz} = \cos z + i\sin z \qquad (2.14.3)$$

I was educated to call this de Moivre's Theorem, though I believe that strictly his Theorem is that,

$$(\cos z + i\sin z)^n = \cos nz + i\sin nz \qquad (2.14.4)$$

a result which follows immediately from (2.14.3) because we now know that $(e^{iz})^n = e^{inz}$ for any integer n, because we have established, following (2.13.7), that $f(z)^n = f(nz)$.

2.15 The Roots of Unity

Having defined $e^{i\theta}$ the reader will now appreciate that any complex number can be written as $z = re^{i\theta}$ for real r and real θ and that this representation is unique in the ranges $r \geq 0, 0 \leq \theta < 2\pi$. These are, of course, simply the polar coordinates in the Argand plane. In particular, complex numbers with modulus unity $(r = 1)$ are $e^{i\theta}$ and we can allow θ, whilst being real, to be unrestricted, i.e., $\theta \in [-\infty, +\infty]$, so long as we recognise that the result is periodic with period 2π.

There are, of course, two square roots of 1, namely 1 and -1. What about the nth root of unity? These are the solutions (roots) of the especially simple polynomial equation $z^n = 1$. But we now know that any n^{th} order polynomial has n complex roots. As we now know that $e^{2\pi i} = 1$ we can write down the n complex nth roots immediately. They are $e^{2\pi i p/n}$ where $p = 0,1,2 \dots (n-1)$. This follows because, despite my earlier words of caution regarding the assumption that $(e^z)^w = e^{zw}$ in general, we know from (2.13.7) that this valid when raising to an integral power. Hence,

$$\left(e^{2\pi i \frac{p}{n}}\right)^n = e^{2\pi i \frac{p}{n} \times n} = e^{2\pi i p} = (e^{2\pi i})^p = 1^p = 1$$

So, for all n possible values of $p = 0,1,2 \dots (n-1)$, $e^{2\pi i p/n}$ is an nth root of unity – and clearly these are all different, so these are all the nth roots of unity – there are no others (within the complex number field). Unity itself is always a root, given by $p = 0$.

On the Argand plane these roots are disposed about the unit circle at equal angular intervals of $2\pi/n$. To drive this home, what we have concluded is that $1^{1/n}$ has n distinct values, 1 being just one of them.

The cube roots of unity are thus $e^{2\pi i p/3}$ where $p = 0,1,2$. $n = 1$ gives what is often denoted $\omega = e^{2\pi i/3}$, whilst the third cube root is then ω^2.

The nth roots of any complex number, $z = re^{i\theta}$, follow immediately as we need only multiply $r^{\frac{1}{n}} e^{i\frac{\theta}{n}}$ by any of the nth roots of unity, where we understand $r^{\frac{1}{n}}$ to refer to the real root. Thus the nth roots of any complex number are,

$$z^{\frac{1}{n}} = r^{\frac{1}{n}} e^{i\left(\frac{\theta + 2\pi p}{n}\right)} \tag{2.15.1}$$

where $p = 0,1,2 \ldots (n-1)$. We see the origin of the trouble we will encounter with exponentiation in the undetermined p that occurs in (2.15.1). That is, raising to a fractional power is a multivalued function. Of course it is, because we know that $w^n = z$ for a given z has n solutions (roots) for w, so $w = z^{\frac{1}{n}}$ has n values because it is a different way of expressing the same fact.

2.16 Deepening Confusion

If we continue to interpret z^a literally, and hence multi-valued in general, things get weird. Suppose q is a prime number, and consider another prime, p, strictly less than q. We now seek the value, or values, of $1^{p/q}$. Because we are here dealing with the sets of all possible values, we are safe to assume that $1^{p/q} = (1^p)^{1/q} = 1^{1/q}$. The set of values taken by $1^{p/q}$ is the same set of values as $1^{1/q}$: $\{e^{2\pi i n/q}, n = 0,1,2 \ldots (q-1)\}$. Consequently, if we attempted to address the problem of the multivalued nature of z^a by interpreting such things as standing for a set of quantities rather than a single quantity, then we would write $1^{p/q} = 1^{1/q}$, meaning the equality of the two sets, despite this being false for individual members of the sets. Algebraic confusion would ensue.

It gets worse. If we let q become arbitrarily large, then the spacing of the roots of $1^{p/q}$ around the unit circle, $2\pi/q$, becomes arbitrarily small. In the limit, "anywhere" on the unit circle is such a root (by which I strictly mean the roots become as dense as we might wish on the unit circle). So, putting $x = p/q$, if we continued to insist that $1^x = (e^{2\pi i})^x$ with a multi-valued meaning, then 1^x could be "anywhere" on the unit circle for arbitrarily large q. Moreover, whilst this is true only when x is the ratio of two different primes, p/q, with $p < q$, any real number, $x < 1$, can be approximated to arbitrary accuracy in this manner (i.e., numbers of the form p/q are dense in the reals).

The same issue would arise when exponentiating any number, real or complex, if the exponent were non-integral. Terms like e^{ikr}, which occur ubiquitously in studying waves of any sort, or in Fourier analysis, would become useless because the phase would not really be kr at all, but completely arbitrary! Mathematical analysis and mathematical physics would collapse in a heap if the multi-valued e^{ikr} were not tamed by making it single valued.

The same will apply, in general, for any z^w where both are complex.

Proof that $\{e^{2\pi i n p/q}, n = 0,1,2 \dots (q-1)\}$ is the same set as $\{e^{2\pi i n/q}, n = 0,1,2 \dots (q-1)\}$

We need to show that the q values of $e^{2\pi i n p/q}$ generated by $n = 0,1,2 \dots (q-1)$ are all different and equal to one the set $\{e^{2\pi i m/q}, m = 0,1,2 \dots (q-1)\}$. Suppose for some $n = n_1$ we write $n_1 p = m_1 q + r$ for some integer m_1 which might be zero, i.e., $r = p(mod\ q)$. The root is thus $e^{2\pi i r/q}$. Proceed by *Reductio ad Absurdum* by assuming there is a second, different, $n = n_2$ also in the range $0,1,2 \dots (q-1)$ which produces the same root, i.e., the same $r = p(mod\ q)$, so $n_2 p = m_2 q + r$ for some integer m_2 which might be zero. So we have $r = n_1 p - m_1 q = n_2 p - m_2 q$ so that,

$$(n_1 - n_2)p = (m_1 - m_2)q$$

But $(n_1 - n_2) < q$ and hence cannot be divisible by q, and neither is p. The above result therefore contradicts the Fundamental Theorem of Arithmetic as it asserts that the LHS has q as a factor, but neither of its terms do – which is impossible. This establishes a contradiction and we conclude that the q values of $e^{2\pi i n p/q}$ generated by $n = 0,1,2 \dots (q-1)$ are all different and hence produce the same set of roots as $\{e^{2\pi i n/q}, n = 0,1,2 \dots (q-1)\}$.

2.17 Logarithms and Riemann Surfaces

Just as we discovered that the apparently sophisticated theory of integration of holomorphic functions was needed to prove something as seemingly simple as the Fundamental Theorem of Algebra, so we shall see that an understanding of logarithms is required before z^u can be defined when both z and u may be complex.

If $z = e^w$, with both quantities complex, then the logarithm of z is defined by $w = log(z)$, i.e., "log" is the inverse of the exponential function (which, recall, is defined by 2.13.3). We know that any given complex w will provide a unique value for $z = e^w$ via that definition. On the other hand, we can add $2\pi n i$ to w, for any integer n, without it changing z. Hence $log(z)$ becomes multivalued for a given z, the values differing by an arbitrary addition of $2\pi n i$.

If $z = re^{i\theta}$ then what is the value of $log(z)$ in terms of r, θ? This can be establish by noting that (2.13.7) is rigorous and hence, if we put $z_1 = e^{w_1}$ and $z_2 = e^{w_2}$, we have $e^{w_1}e^{w_2} = z_1z_2 = e^{w_1+w_2}$, and hence that,

$$log(z_1z_2) = w_1 + w_2 = log(z_1) + log(z_2) \qquad (2.17.1)$$

So we deduce that,

$$w = log(z) = log(re^{i\theta}) = log(r) + log(e^{i\theta}) = log(r) + i\theta \quad (2.17.2)$$

The problem is that, given z, we know θ only up to an arbitrary addition of $2\pi n i$ for some integer n. The same z could be written $re^{i(\theta+2\pi n)}$ but this would give a different $log(z) = log(r) + i(\theta + 2\pi n)$.

Whether it be $log(z)$ or $z^{1/q}$, some orderliness can be brought to these multivalued functions via the concept of the Riemann surface. A function $f(z)$ is multivalued if the mapping $z \rightarrow f(z)$ is one-to-many. To a single point in the domain (the Argand plane of z) there are many values in the range of the function. A trick to turn this into a 1-to-1 (bijective) mapping is to change the domain from the Argand plane to a Riemann surface, which can be thought of as a set of parallel Argand "planes" joined together in a topologically non-trivial manner. Thus, we consider each value of $f(z)$ for a given z to arise from a different "layer" of the Riemann surface.

Figures 2.3 and 2.4 illustrate the Riemann surface for $\sqrt{z}$ and $log(z)$ respectively (the surface has been "pulled" in the normal direction to separate its layers to aid visualisation). The former has just two "layers", i.e., two points on the Riemann surface for a given z. The latter has infinitely many, though only three layers are shown. In both cases the different "layers" of the Riemann surface correspond to different 2π ranges of the argument of z. In Figure 2.3 for $\sqrt{z}$ the value of θ where the surface self-intersects is arbitrary. In the illustration the intersection is at $\theta = \pm\pi$ and hence lies on the negative real axis. It could equally be placed on the positive real axis, $\theta = 0$ or 2π.

Figure 2.3: Example of a Riemann Surface (this for $\sqrt{z}$). By Leonid 2 - Own work, CC BY-SA 3.0. https://commons.wikimedia.org/w/index.php?curid=38245841

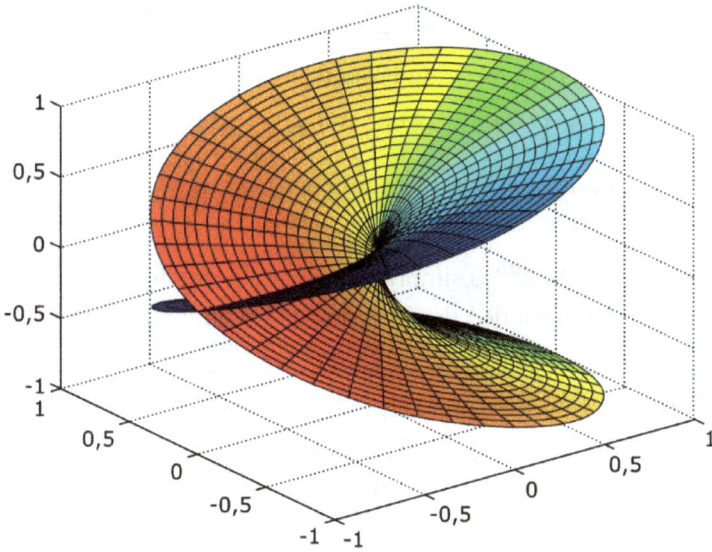

Figure 2.4: Example of a Riemann Surface (this for $\log (z)$). By Leonid 2 - Own work, CC BY-SA 3.0, https://commons.wikimedia.org/w/index.php?curid=38074912

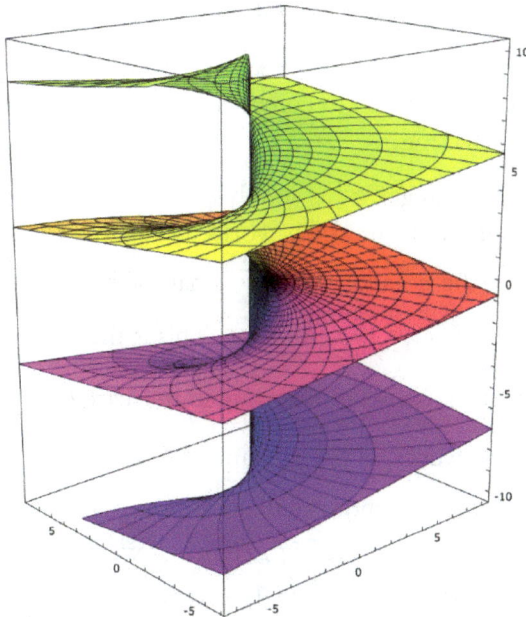

2.18 The Trouble with Exponentiation

Let's start at 1827 with the following paradox noted by Thomas Clausen (see Steiner et al), though I use a simpler variant,

We know from (2.14.3) that $e^{2\pi i n} = 1$ if n is an integer. If we uncritically assumed that $1^z = 1$ for any z and also that $(e^z)^w \equiv e^{zw}$ then we get this,

$$1 = 1^{2\pi i n} = (e^{2\pi i n})^{2\pi i n} = e^{-4\pi^2 n^2} \qquad (2.18.1)$$

which is clearly nonsense. What has gone wrong can be expressed in several different ways, all related,

- Actually $(z^a)^b \equiv z^{ab}$ is simply not generally true for complex numbers, dumbfounded though you will no doubt be to hear it.

- In order to even begin to consider whether two functions of z might be identical, both must be uniquely defined. And yet, z^a, where a is the reciprocal of an integer, $a = 1/n$, is not uniquely defined but can take any one of n values. So how can any claimed identity involve z^a?

- The notation of the exponential functions as e^z is asking for trouble because, despite appearances, it does not really mean "raise the number e to the power z". Actually e^z is defined as the function given by (2.13.3). The difference is crucial, as (2.13.3) always yields a unique value – in contrast to the literal interpretation of e^z which becomes multi-valued for fractional z.

Let's start simply and work up to complex numbers "raised to a complex power", whatever that means.

When n is a positive integer, z^n is simply a compact notation for $z \times z \times ... z$ with n factors (and this works just as well for complex as real z). If $z^n = w$ then the notation $z = w^{\frac{1}{n}}$ is natural because then we have $w^{\frac{1}{n}} \times w^{\frac{1}{n}} \times ... w^{\frac{1}{n}} = \left(w^{\frac{1}{n}}\right)^n = z^n = w$. This is the origin of the identification of $(w^a)^b$ with w^{ab} because in this case $\left(w^{\frac{1}{n}}\right)^n = w^{\frac{1}{n} \times n} = w$ is clearly correct. However, as things stand, $w^{\frac{1}{n}}$ is not a unique quantity.

Negative exponents can be motivated in a similar manner. By definition, for integers n, m we have $z^n z^m \equiv z^{n+m}$. Hence, if we write $w = \frac{1}{z^n} = z^{-n}$ then $wz^n = z^{-n}z^n = z^{n-n} = z^0 = 1$, as required. Combining the two notations shows that $z^{-\left(\frac{1}{n}\right)}$ is also a consistent notation, though ambiguous without further clarification.

Leaving aside for a moment its multi-valued nature, the above notational definitions mean we can attach meaning to $z^{p/q}$ for complex z and real integral p and q (positive or negative). Approaching irrational numbers as the limit of very large p and q this suggests that z^x for any real x and y will respect the familiar identities $z^x z^y = z^{x+y}$ and $(z^x)^y = z^{xy}$.

However, one must accept that, as yet, this is all ill-defined because we are still dealing with multi-valued quantities, so any claimed identities are treacherous because ambiguous. In the case that the so-called "base", that is the z in z^x is real, and x is also real, then we can simply adopt the convention that, unless stated to the contrary, in algebraic workings we understand z^x to be, out of the potentially many possible quantities, the only real quantity.

Moreover, because the above approach to y^x for real but irrational x was rather heuristic, we can make it better defined thus (for real x and real y),

$$y^x \equiv e^{x\log(y)} \qquad (2.18.2)$$

This is the generalisation of $\log(y^n) = n\log(y)$ for integer n. Here it is understood for real y that $\log(y)$ is real, i.e., we take the value from the $n = 0$ layer of the Riemann surface. Based on (2.13.7) this allows us consistently to define a real number exponentiated by a complex exponent, thus, $\qquad (2.18.3)$

$$a^{x+iy} = a^x a^{iy} = a^x e^{iy\log(a)} = a^x\big(\cos\big(y\log(a)\big) + i\sin\big(y\log(a)\big)\big)$$

where the a^x part is strictly defined by (2.18.2) and the a^{iy} part follows from (2.18.2) combined with de Moivre, (2.14.3). Note that a^{x+iy} becomes uniquely defined by (2.18.3) only because a is real and it is understood therefore that $\log(a)$ takes its real value (on the $n = 0$ layer of the Riemann surface).

Note that this does not allow us to examine whether $(a^z)^w$ is the same as a^{zw}, even for real a, if z is complex, because a^z is then complex, in general, and we have provided no definition of the exponentiation of a complex number. However, this familiar identity does hold if w is a real integer, which the reader can readily prove using (2.14.4). It also holds when z is real, for arbitrary complex w. The reader can also readily prove this based on the validity of the familiar identities for real quantities: $(a^b)^c = a^{bc}$ and $log(a^b) = b \cdot log(a)$.

That leaves us with the thorny issue of the exponentiation of a complex number (whether the exponent is real or complex). Lamentably, this remains a messy business in which there is no unique definition which will permit you to use the familiar identities when the base is complex, that is, in general for complex z, $(z^b)^c \neq z^{bc}$ and $log(z^b) \neq b \cdot log(z)$.

We cannot even adopt an approach in which multi-valued quantities are understood to stand for the set of all its values, because these identities even fail as an equality between sets. For example, for a positive integer q,

$$log(z^q) = log\big((re^{i\theta})^q\big) = log(r^q) + log\big(e^{iq\theta}\big)$$

$$= q \cdot log(r) + i(q\theta + 2\pi n) \qquad (2.18.4)$$

Whereas, $\qquad\qquad\qquad\qquad\qquad\qquad\qquad\qquad\qquad\qquad (2.18.5)$

$$q \cdot log(z) = q(log(r) + i\theta + 2\pi ni) = q \cdot log(r) + i(q\theta + 2\pi nq)$$

To drive home the difference between the set of values specified by (2.18.4) and that given by (2.18.5), consider $q = 5$. The imaginary parts allowed by (2.18.4) are $\{... 5\theta - 4\pi, 5\theta - 2\pi, 5\theta, 5\theta + 2\pi, 5\theta + 4\pi ...\}$ compared with $\{... 5(\theta - 4\pi), 5(\theta - 2\pi), 5\theta, 5(\theta + 2\pi), 5(\theta + 4\pi) ...\}$ allowed by (2.18.5), which are quite different.

One might be tempted to think that we could adopt a convention of insisting that $log(z)$ is always assigned an imaginary part which lies in a given 2π range, say $[0, 2\pi)$. The result is called the Principal Value of $log(z)$. But that fails to salvage the identity and in general we still find $log(z^b) \neq b \cdot log(z)$. For real b the imaginary part of $log(z^b)$ is then in the range $[0, 2\pi)$ by definition, whilst that of $b \cdot log(z)$ need not be. For example, putting $z = e^{i5\pi/4}$ and $b = 2$, using Principal Values we have $log(z^b) = i\pi/2$ whilst $b \cdot log(z) = 5\pi i/2$.

Similarly, we can define a unique exponentiation of a complex number by,

$$z^w = e^{wlog(z)} \qquad (2.18.6)$$

where both z and w might be complex, and make this single-valued by use of the Principal Value of the logarithm. But we shall now find $(z^w)^v \neq z^{wv}$ in general. In fact we find,

$$(z^w)^v = z^{wv} e^{2\pi n v i}$$

where the integer n is determined by setting $log(e^{w \cdot log(z)})$ to its Principal Value. For example, if $z = e^{i5\pi/4}$ and $w = 2$ then the Principal Value of $log(e^{w \cdot log(z)}) = log(e^{i5\pi/2})$, namely $i\pi/2$, is obtained by setting $n = -1$. Thus, if $v = 1/\pi$, taking the Principal Value route to unique quantities leads to $(z^w)^v$ and z^{wv} differing by a factor of $e^{-2i} = -isin(2)$.

Thus, you should be extremely cautious about assuming the identities $(z^b)^c = z^{bc}$ and $log(z^b) = b \cdot log(z)$, familiar from real algebra, when z is complex. In general they are not true. This is the explanation of the paradox of (2.18.1). The identity $(z^b)^c = z^{bc}$ was assumed in simplifying $(e^{2\pi i n})^{2\pi i n}$, but it just is not valid.

2.19 References

Steiner, J.; Clausen, T.; Abel, N. H. (1827). "Aufgaben und Lehrsätze, erstere aufzulösen, letztere zu beweisen" [Problems and propositions, the former to solve, the later to prove]. Journal für die reine und angewandte Mathematik. 2: 286–287. Republished by De Gruyter December 14, 2009, https://doi.org/10.1515/crll.1827.2.286.

3

The Square Root of Probability

Wherein we enter the world of quantum physics, but not before finding that the concept of probability was problematical even before the quantum physicists got their hands on it. But only in quantum mechanics do we encounter the square root of probability. Remarkably, the resulting probability amplitudes, when combined with Feynman's sum-over-paths, allows the deduction of a range of fundamental features of physics, including Newton's First Law, the conservation of energy and momentum, and the fact that particles have a well-defined mass.

3.1 Probability – Slippery and Clean

If ever there were a slippery concept, it is probability.

The following discussion orbits around the objective versus the subjective aspects of probability, before landing upon the author's personal perspective on that issue in the context of quantum mechanics. It may not be what you expect.

Even when a probability is clear and certain, our intuition often leads us astray. Let me illustrate. Consider this scenario: a man you do not know tells you he has two children and he happens to mention that one of the children is a boy. What is the probability that the man's other child is also a boy? (For the purposes of this question, ignore the possibility of twins and assume girls and boys are equally likely).

Even strong mathematicians get the answer wrong. Most people will answer 50% (after pausing, suspecting a trick question). But the answer is 1/3 (~33%). Intuition misleads as follows: the second child is independent of the first, and any birth has a 50% chance of being a boy – therefore the answer is 50%, apparently - but it isn't.

My personal Golden Rule is that, whenever possible, probabilities should be defined and quantified by a ratio of numbers of events. Count the possibilities! Imagine the above scenario being played out, not once, but many times. You can simulate this on a computer. The man in question is,

(a) assigned a first child at random, either M or F (50/50);

(b) assigned a second child at random, independently of the first, and hence again M or F (50/50).

Clearly, four permutations occur: MM, MF, FM and FF. Over a very large number of simulations each of these four possibilities will occur nearly the same number of times (the fractional difference is $\sim 1/\sqrt{N}$ which tends to zero for large N). We are told only that one of the children is a boy, so that rules out only the case FF. We are left with three cases: MM, MF, FM, of which only one has the other child being a boy, ergo a probability, defined by the relative frequency of occurrence, of $1/3$.

Even after this is explained to people, I have found they still argue – sometimes saying "it's a matter of opinion". No, it isn't. It is clear and certain. It is people's grasp of probability which is not clear or certain.

Having hammered home that lesson about enumerating possibilities, here are three variants on the initial question – in each case you need to evaluate the probability that the man's other child is a boy…

Version 2: A man you do not know tells you that he has two children and the elder child is a boy.

Version 3: A man you do not know tells you that he has two children and one of the children is a boy born on a Tuesday.

Version 4: A man you do not know tells you that he has two children and the elder child is a boy born on a Tuesday.

The answers are,

Version 2: $1/2$

Version 3: $13/27$

Version 4: $1/2$

It is hard to believe that being told that the boy in question is the elder child can change the probability that the other child is a boy. But it does. Of the four initial permutations, MM, MF, FM, FF, interpreting the first to be the first-born child, in Version 2 both FM and FF are now eliminated leaving just MM and MF, hence the probability of the other child being a boy is $1/2$.

And even when you have grasped that, your mind will balk at the idea that specifying the day of the week on which one child is born can have any bearing on the sex of the other. Yet it does. I'll leave versions 3 and 4 as an exercise. (Hint: there are now 14 types of child, two sexes times 7 birth days, and $13 = 14 - 1$ and $27 = 2 \times 14 - 1$).

The serious failure of intuition to get probabilities right can have enormous commercial implications, for example in engineering. Consider a case in which there is a very large number of nominally identical components (N) in some engineered facility. Each component has only a small probability of failure ($P \ll 1$), but nevertheless there are failures occasionally simply due to the large number of components (PN is approaching unity). Suppose a few components (say 5 of them) are subject to an unusually onerous abnormal condition, such as running at an abnormally high temperature. Suppose it is known that this increases the probability that these components will fail by an order of magnitude (so, to $10P$ per item). Traditionally-minded engineers will be inclined to put all their efforts into addressing the abnormal operating temperatures of these 5 components – even when $5 \times 10P \ll (N - 5)P$, in other words even when it is far more likely that one of the $N - 5$ well-behaved components will fail rather than one of the 5 hot components – simply due to the former's far greater numerical preponderance. Even when operating records confirm that it has never been hot components that have failed in the past, only ever components at normal temperature – still engineers will tend to cling to their conviction that something must be done about the hot components, possibly spending a great deal of money to fix the 'problem'. (Yes, this is a real-life example from my own experience – and if one's fellow engineers are hard to convince, you can imagine the problem with managers).

So, the Golden Rule is "count the possibilities". This is known as the "frequentist" or "physical" interpretation of probability. Whenever it is possible to deploy this method, then do so. In practice this might mean reproducing a large number of events with as near to identical pre-conditions as one can achieve.

An example might be estimating the probability that a given engineering component will fail under given operating conditions. In principle this could be established by manufacturing many identical components and subjecting them to the required conditions in a test centre. There is, however, a huge difference between principle and practice is such cases – because large engineering components are costly to make. Whilst a few might be made specifically to test, perhaps to destruction, not many will. And the number needed to estimate empirically a small failure probability, P, is larger than $1/P$ and hence not a practical or economic possibility. In practice, the probability of engineering failures is estimated by calculation not by testing.

This is a more significant shift of methodology than merely an avoidance of cost. It replaces the "frequentist" quantification of probability with a knowledge-based approach. A calculation of failure probability can be carried out only because we have a substantial body of knowledge about failure mechanisms and their quantification, as well as knowledge about the statistical distributions of the material properties, loads and operating conditions which enter the calculation.

Is the "frequentist" quantification of probability always applicable, if only in principle? What about, say, the probability it will rain in Gloucester tomorrow (written when "today" was 23rd July 2021)? We cannot replay the past to re-run the weather on 24th July 2021 in Gloucester a large number of times. Yet we are still happy to attach a meaning to "the probability it will rain in Gloucester tomorrow". There is a degree of faith in assuming such a thing is meaningful. That we can, by various means, make estimates of the supposed quantity enhances confidence that the quantity exists – but should it?

For example, we could make a first estimate of the probability by dividing the number of rainy days recorded last year by 365. We would find it necessary to be specific about location, as Rannoch Moor will not be indicative of Gloucester (despite Dr Foster's unfortunate experience). Further refinement would be required for passable accuracy by concentrating on past rainy Gloucester days specifically in July, and dividing by 31. By such familiar means an empirical "frequentist" estimate is obtained for the probability of a one-off event. Implicitly, the claim is that tomorrow is another day drawn from the same statistical population as other Gloucester days in July. On this basis we find that the probability of rain tomorrow in Gloucester is 23%. And because we can quantify it, we are happy that there exists an objective something which we refer to as "the probability of rain tomorrow in Gloucester".

Happy? You should not be. Because I have just looked at the Meteorological Office's forecast for tomorrow and the prediction is 90% for heavy rain. The Met Office is not infallible (ha!) but, coming after 10 days of very high temperatures for this area, circa 30°C, and noting the falling atmospheric pressure, I am inclined to believe it. (After the event I can tell you they were correct – the rain came heavily).

So, the probability of rain in Gloucester on a random day in July is 23% except when we know better. Hmm. When do we know better? Whenever our broadband is working and the Met Office is not on strike? These are odd things for the weather in Gloucester to depend upon. You see how slippery is this concept of probability in practice? It is no wonder many mathematicians do not regard probability theory as proper maths. It does seem that the concept of probability is hopelessly contaminated with the subjective, namely the state of knowledge of the person making the estimate of probability.

People who work in the field of probabilistic engineering assessments often make a distinction between aleatory and epistemic uncertainties. An aleatory uncertainty is conceived to be an irreducible, unavoidable uncertainty – the word "aleatory" being derived from the Latin "alea" meaning a die, or the rolling of dice. In contrast, epistemic uncertainties are due to our ignorance: in principle an epistemic uncertainty could be reduced or eliminated if we had more information.

My personal view is that the distinction between aleatory and epistemic uncertainties is rather arbitrary. In a deterministic world, in which perfect present knowledge would permit a perfectly certain prediction of the future, there would be no aleatory uncertainties. All uncertainties would be attributable to our lack of knowledge. Even the dice which give "aleatory" its name would, in principle, be predictable if we knew the initial "boundary conditions" of their motion with sufficiently accuracy, and, of course, perfect knowledge of air currents and the exact topography and frictional behaviour of the surface upon which they landed.

Actually, of course, even in the classical Newtonian world, chaotic behaviour renders almost everything unpredictable. Weather is a case in point. Weather is chaotic, which means that prediction beyond a certain number of days is not possible, however much satellite data you possess to define current conditions and however expensive your supercomputer. This gives us a practical answer to the conundrum of the probability of rain in Gloucester. Over the short term, the Met Office is the best estimate, whilst beyond a certain number of days, the frequentist empirical estimate is all we can do. (The number of days ahead for which the Met Office provides the better estimate is somewhere between 7 and 10 days, and 14 days may be the ultimate limit).

This brief discussion of the vagaries of probability, and in particular the failure of intuition as a reliable guide, would not be complete without mention of

Bayes: his followers and his Theorem. One needs to distinguish Bayes Theorem, which is beyond question, from the interpretation which those who call themselves Bayesians put upon probability. A discussion of Bayes' Theorem I relegate to Appendix I.

Bayesians are the counterpoint to frequentists. They are happy to regard probability as being a subjective issue. Probability, they argue, reflects a personal belief or state of knowledge of an individual. (Exit mathematicians in disgust, stage left). They have a point, as the above examples illustrate. Forced to estimate any probability, what we come up with depends crucially on our state of knowledge (such as knowing what the Met Office is saying, or the accuracy of our understanding of structural failure). The "man with two children" problems also illustrate very clearly how probability depends upon what we know. Whether a child is the eldest or born on a Tuesday can have a surprising influence on the probability of the other child being male, rather bewilderingly.

You can think of any number of such examples. Here's another. What is the probability that I will die of a certain cancer? I can take the empiricist-frequentist course again, looking at the statistics and, if the data are available, filtering it for: (a) males, (b) males of my age, (c) males of my age and ethnicity, (d) males of my age, ethnicity and dietary habits, (e) males of my age, ethnicity, dietary habits and body-mass index, (f) males of my age, ethnicity, dietary habits, body-mass index and history of exercise, etc. etc. It seems I could home in on an accurate probability.

But what if I did not know that this type of cancer was always associated with a specific gene? And I did not know that I do not have that gene? My previous, laborious, statistical researches were pointless. I'm safe from this particular threat; the probability is zero.

My own spin on this example is that, whilst our best estimate of the probability is clearly dependent upon our state of knowledge, one could argue that a well-defined probability exists (zero, in this case) even if we are ignorant of the genetic issues and hence fail to accurately estimate it. But even that article of faith fails to give us reassurance that, in any given situation, our estimate of probability is even approximately equal to this supposed objectively existent ideal. Since probability does seem irreducibly to depend upon our state of knowledge, how can we ever be sure that there is not something extra to know that would change our estimate radically?

Information and entropy are closely connected with probability. Information, many people these days will insist, is physical. Yet there are problems quantifying information, which relate to knowledge, and over which a veil tends to be drawn. A string of N binary digits only constitutes N bits of information if there is no pattern in the sequence that we happen to recognise. If they are all 0s, or have an obvious pattern like 01010101 etc., then however long the string of digits we can agree there is no more than a bit or two of information. In other cases the sequence may look random, and hence genuinely to contain N bits of information – until some genius points out that it is just π, in binary. How much information is there in π to 1000 binary digits? Is there any more information in one million digits? If you adopt an algorithmic definition of information, then no.

Entropy always appears as a physical quantity, and hence objective. But this may be because we use entropy only in situations which are sufficiently well-defined that the ambiguities are avoided. There is no all-knowing entity – no kinetic theory Met Office - who suddenly gifts us with a knowledge of the velocities of all the molecules in a gas. So the inherent fraudulence of our usual frequentist approach (namely, statistical thermodynamics) is safe from exposure by some omniscient Maxwell demon only because the latter does not exist.

All these ambiguities in estimating probabilities relate to whether there is more information potentially available than we currently possess. In other words, the problem lies in the epistemic uncertainties. By definition, these are uncertainties about which we might be able to obtain additional information. This additional information will change our estimate of the corresponding probabilities – perhaps radically. So, all the ambiguities in estimating probabilities can be laid at the door of the epistemic uncertainties. If all the uncertainties were aleatory, i.e., irreducibly uncertain and about which no additional information could ever be obtained even in principle, then all the vagaries in probability would vanish.

And so to quantum mechanics.

Leaving aside chaos, in a fully deterministic physics, actually there are no truly aleatory uncertainties. In principle all uncertainties are epistemic (though the gulf between principle and practice may be unbridgeable). And probabilities estimated in any situation with epistemic uncertainties will be subject to ambiguity, i.e., sensitivity to potential additional information.

But there is one type of physical system in which uncertainty is genuinely and irreducibly aleatory – pure quantum systems. Measurements on pure quantum states have probabilistically determined outcomes, but the uncertainty in which outcome will be obtained is purely aleatory. It is not possible, even in principle, to obtain additional information which could render the measurement outcome determinate.

Physical reality includes hybrid states which display some 'quantum' behaviour, with its aleatory uncertainties, but also some 'classical' behaviour, which will generally involve epistemic uncertainties. But there are special systems of particular interest which are in **pure** quantum states. These states are subject to aleatory uncertainty alone. A system is in a pure quantum state if all the epistemic uncertainties have been honed out of it.

In that sense, then, the issue of probability is cleaner for pure quantum states than for classical situations. The ambiguities discussed in this section abound in everyday applications of probability, but pure quantum states do not share this same ambiguity.

That is not because probability is irrelevant in quantum physics – on the contrary, quantum mechanics is essentially probabilistic. It is because pure quantum states are purely aleatory and free of epistemic uncertainty, as a result of which the probabilities in question are well-defined and free of the types of ambiguity discussed in this section.

This is rather fortunate for physics, isn't it?

Many words have been devoted to the issue of the role of an observer's knowledge (or consciousness) in determining the outcome of a measurement on a quantum system. The interpretational position adopted here is that there is none. It there were, epistemic issues would remain and the probabilities of quantum mechanics would be alterable upon the attainment of additional information. The assertion that pure quantum systems are purely aleatory saves the probabilities of quantum mechanics from being subject to the ambiguities that plague probabilities in other (classical or everyday) settings.

It is the irreducibly indeterminate nature of quantum measurements on pure states that makes probability in quantum mechanics well-defined. Indeterminism is, in that sense, our friend.

One hears people assert that quantum mechanics is contaminated by the subjective, that our state of knowledge has a bearing on the outcome of measurements – and that this contrasts with classical physics which is strictly objective. Well, classical physics *is* strictly objective, but probability in the classical, or everyday, context is not. As we have seen it is subject to our state of knowledge. It is ironic, then, that pure quantum states are the one case where probability attains its fully objective status: for pure quantum states probability is clean, not slippery as it is in everyday usage.

3.2 The Origin of Probability in Quantum Mechanics

I will defer a description of the formulation of quantum mechanics to following sections and chapters. In this section I reprise the history of how mechanics, which in its classical Newtonian form was the very essence of determinism, gave way to a new form of mechanics which was indeterminate and unavoidably probabilistic. The cognitive pain engendered in the physicists of the day by this conceptual revolution should not be underestimated.

The man credited with the first cogent interpretation of the emerging quantum mechanics in probabilistic terms was Max Born. The convoluted story involving many actors is sometimes bowdlerised by referring to Schrodinger's 1926 paper (which presented his famous equation) and noting that, within days, Born had interpreted Schrodinger's otherwise obscure psi function (ψ) probabilistically. Specifically, Born identified $|\psi|^2$ with the particle's probability density in space. However, Born was able to make this interpretation so quickly only because he had taken the key conceptual steps the previous year.

The achievements of physicists in the first three decades of the twentieth century were quite staggering. At the turn of the century it was still possible for informed physicists to doubt the existence of atoms. Whilst chemistry and the kinetic theory of gases pointed to the existence of atoms or molecules, it was Einstein's 1905 work on Brownian motion that clinched the matter. The first 28 years of the twentieth century saw the rise of special relativity, general relativity, quantum mechanics, Dirac's relativistic quantum equation and Dirac's prediction of antimatter – and, by 1932/3, the observation of antimatter. In these three decades our understanding of space and time was radically altered, and mechanical determinism, which had ruled science since the time of Newton, was overturned.

Anyone asked to name half a dozen or so of the most significant contributors to the start of quantum mechanics would be certain to include Planck, Einstein, De Broglie, and Schrödinger. Yet all four of those great physicists remained dissatisfied all their lives with the new quantum mechanics, and specifically its indeterminate or probabilistic nature. Schrodinger introduced his famous equation in the hope that it would reintroduce determinism into the emerging new mechanics applicable to atoms. But the Born interpretation of the wavefunction (i.e., ψ) was to put paid to that hope.

In the mid-1920s, there were three centres energetically developing quantum mechanics: in Germany, Switzerland and England. At the University of Göttingen there was Born, Heisenberg and Jordan, of which Born was the senior man (the others being his students). Schrodinger was employed by the University of Zurich in 1921 to 1926, though he was in a Swiss tuberculosis sanatorium when he developed his famous equation. And in Cambridge, England, there was Paul Dirac. The development of quantum mechanics was a complex interplay between these men, with key contributions also from Einstein and Pauli. Bohr's earlier work was the springboard for the further developments during this period.

Heisenberg was awarded the Nobel Prize for his work on quantum mechanics in 1932, and Schrodinger and Dirac won the prize in 1933. Oddly, Born did not share either of these prizes, despite Einstein nominating all three of the Göttingen men (rather noble of him, given that he didn't believe much of it). The omission of Born is especially odd as, not only did Born introduce the probabilistic interpretation, and he was the senior man, but he was also the first to explicitly write the canonical commutation relation in a publication (in 1925). However, this omission was corrected – rather belatedly – in 1954.

For the rest of this brief history I will quote from Max Born's 1954 Nobel Prize speech (extracts only), which focusses specifically on the origin of the probabilistic interpretation of quantum mechanics (quotes in italics),

"At the beginning of the twenties, every physicist, I think, was convinced that Planck's quantum hypothesis was correct. According to this theory energy appears in finite quanta of magnitude $h\upsilon$ *in oscillatory processes having a specific frequency* υ *(e.g., in light waves). Countless experiments could be explained in this way and always gave the same value of Planck's constant,* h*"*

"*Again, Einstein's assertion that light quanta have momentum* $h\upsilon/c$ *(where* c *is the speed of light) was well supported by experiment (e.g., through the Compton effect). This implied a revival of the corpuscular theory of light for a certain complex of phenomena. The wave theory still held good for other processes. Physicists grew accustomed to this duality and learned how to cope with it to a certain extent. In 1913 Niels Bohr had solved the riddle of line spectra by means of the quantum theory and had thereby explained broadly the amazing stability of the atoms, the structure of their electronic shells, and the Periodic System of the elements. For what was to come later, the most important assumption of his teaching was this: an atomic system cannot exist in all mechanically possible states, forming a continuum, but in a series of discrete 'stationary' states. In a transition from one to another, the difference in energy* $E_m - E_n$ *is emitted or absorbed as a light quantum* $h\upsilon$ *(according to whether* E_m *is greater or less than* E_n*). This is an interpretation in terms of energy of the fundamental law of spectroscopy discovered some years before by W. Ritz.*"

"*The situation can be taken in at a glance by writing the energy levels of the stationary states twice over, horizontally and vertically. This produces a square array,*

	E1	E2	E3
E1	11	12	13
E2	21	22	23
E3	31	32	33

in which positions on a diagonal correspond to states, and non-diagonal positions correspond to transitions."

"*The problem was this: an harmonic oscillation not only has a frequency, but also an intensity. For each transition in the array there must be a corresponding intensity. … the decisive step was again taken by Einstein who, by a fresh derivation of Planck's radiation formula, made it transparently clear that the classical concept of intensity of radiation must be replaced by the statistical concept of transition probability. To each place in our pattern or array there belongs (together with the frequency* $\upsilon_{mn} = (E_m - E_n)/h$*) a definite probability for the transition coupled with emission or absorption.*"

It was Born who gave us the term "quantum mechanics",

"*Investigations into the scattering and dispersion of light showed that Einstein's conception of transition probability as a measure of the strength of an oscillation did not meet the case, and the idea of an amplitude of oscillation associated with each transition was indispensable. …. A paper of mine, which introduced, for the first time I think, the expression quantum mechanics in its title…*"

Remarkable as it seems now, matrices were not a standard part of scientist's toolkit in the 1920s, so their use in quantum mechanics was quite avant-garde,

"I could not take my mind off Heisenberg's multiplication rule, and after a week of intensive thought and trial I suddenly remembered an algebraic theory which I had learned from my teacher, Professor Rosanes, in Breslau. Such square arrays are well known to mathematicians and, in conjunction with a specific rule for multiplication, are called matrices. I applied this rule to Heisenberg's quantum condition and found that this agreed in the diagonal terms. It was easy to guess what the remaining quantities must be, namely, zero; and at once there stood before me the peculiar formula,

$$pq - qp = h/2\pi i$$

This meant that coordinates q and momenta p cannot be represented by figure values but by symbols, the product of which depends upon the order of multiplication - they are said to be 'non-commuting'."

Note that Born's paper with the above canonical commutation relation was in 1925. His account continues,

"The first non-trivial and physically important application of quantum mechanics was made shortly afterwards by W. Pauli who calculated the stationary energy values of the hydrogen atom by means of the matrix method and found complete agreement with Bohr's formulae. From this moment onwards there could no longer be any doubt about the correctness of the theory."

Pauli submitted his epochal paper deriving the energy levels of hydrogen using the algebraic method on 17th January 1926. Schrodinger submitted his derivation of the same result using his wave equation just 10 days later. And just a few days after the publication of the latter, Born realised that Schrodinger's wave function should have a probabilistic interpretation. In contrast, Born wrote in 1954,

"Schrödinger thought that his wave theory made it possible to return to deterministic classical physics."

Born explains his realisation that Schrodinger's wavefunction must be interpreted probabilistically as follows,

"Again an idea of Einstein's gave me the lead. He had tried to make the duality of particles - light quanta or photons - and waves comprehensible by interpreting the square of the optical wave amplitudes as probability density for the occurrence of photons. This concept could at

once be carried over to the ψ-function: |ψ|² ought to represent the probability density for electrons (or other particles)."

Regarding the acceptance of the probabilistic interpretation by the physics community, Born wrote,

"However, a paper by Heisenberg (1927), containing his celebrated uncertainty relationship, contributed more than the above-mentioned successes to the swift acceptance of the statistical interpretation of the ψ-function. It was through this paper that the revolutionary character of the new conception became clear. It showed that not only the determinism of classical physics must be abandoned, but also the naive concept of reality which looked upon the particles of atomic physics as if they were very small grains of sand. At every instant a grain of sand has a definite position and velocity. This is not the case with an electron. If its position is determined with increasing accuracy, the possibility of ascertaining the velocity becomes less and vice-versa."

Of particular pertinence to the manner in which "probability" is manifest experimentally is this remark of Born's,

"Almost all experiments lead to statements about relative frequencies of events, even when they occur concealed under such names as effective cross section or the like."

Consequently, the use and interpretation of probability in quantum mechanical experiments is generally the straight-forward "frequentist" interpretation. The same cannot be said, however, for its square root – the wavefunction, ψ .

3.3 The Square Root of Probability: Quantum Mechanics

Let me confess immediately that I am taking a liberty with the idea of "square root" in this example. Normally the square root, ψ, of a quantity, P, would be such that $\psi^2 = P$. Here, though, we wish the "square root" entity, ψ, to be complex, whilst the probability, P, is necessarily real. To ensure this we write instead,

$$|\psi|^2 = P$$

which means that,

$$\psi = e^{i\theta}\sqrt{P} \tag{3.3.1}$$

So the quantity ψ (let us call it the "wavefunction") is the square root of probability times a "phase factor", that is a complex number of modulus unity, namely $e^{i\theta}$ for real θ. The immediate corollary of this is that, whilst the probability can be uniquely derived from the wave function, ψ, the reverse is not true. Knowing the probability does not uniquely define the wavefunction, but only does so up to some phase factor.

Probability of what, I hear you reasonably ask? If it is a particle we are talking about, then $P = |\psi|^2$ would actually be interpreted, not as a simple probability, but as a probability density, namely the probability per unit spatial volume that the particle occurs at the point in question. However, we might be more interested in its momentum, in which case P might be the probability of a certain momentum, per unit range of momentum. Or we might be interested in the spin (or other angular momentum) of the particle, in which case P will be the probability of the spin taking some chosen value. Or we might be interested in the transition of a system from one state to another, in which case P would be the probability of that transition, given certain starting conditions. In that latter case, the ψ appearing in $P = |\psi|^2$ would not be called a wavefunction but a transition amplitude or (for reasons that will be made clear below) a matrix element.

The wavefunction's phase factor is a curious thing. It is both irrelevant and crucial. For a single wavefunction the phase factor is physically irrelevant: it is unobservable. Measurement will only observe the probability, P, from which the phase factor vanishes (or the probabilities of measurable physical quantities such as position, momentum, energy, etc., which are again independent of the phase).

But when two or more wavefunctions combine (i.e., physically interact), their relative phase is crucial. It is their relative phase which determines whether the wavefunctions interfere constructively or destructively. It is the phase factor which provides the wavefunction with its "wave" characteristics. To complete this picture we need to understand how the phase angle, θ, varies in space. To rationalise this, let us reprise the history briefly.

3.3.1 Probability Amplitude

The wavefunction, ψ, is a probability amplitude – a concept that simply did not occur in classical physics – nor, for that matter, in probability theory either prior to the advent of quantum mechanics. A probability amplitude is a quantity which stands in relation to a probability as given by (3.3.1), that is, as a sort of generalised, complex-valued square root such that $|\psi|^2 = P$. Curiously, and crucially, probability amplitudes have the same two laws of composition as classical probabilities. These are,

[1] If *independent* events A, B, C,… have separate probability amplitudes $\psi_A, \psi_B, \psi_C, …$ then the probability amplitude for all those events occurring (perhaps one after the other) is the product of the amplitudes,

$$\psi_{total} = \psi_A \psi_B \psi_C …$$

[2] If events A, B, C,… are *alternative* events with the same outcome, X (either A occurs or B occurs or C occurs, etc., and all possibilities are covered) and these alternatives have individual probability amplitudes $\psi_A, \psi_B, \psi_C, …$ then the overall probability amplitude for X is the sum of the amplitudes,

$$\psi_X = \psi_A + \psi_B + \psi_C + \cdots$$

In other words: logical "and" = product, whilst logical "or" = addition, just as it does for classical probabilities. But these rules of composition, specifically the addition rule for probability amplitudes, means that probabilities in quantum mechanics behave differently from classical probabilities.

On forming the square modulus of the product rule for "and", [1], we get,

$$P_{total} = P_A P_B P_C …$$

which is the usual rule for the probability of all the events A,B,C,… occurring, assuming they are independent. But the corresponding conclusion for the addition rule does not hold.

The addition rule is also called the "superposition of wavefunctions" or "the superposition principle". From the perspective of a wave equation, $\hat{Z}\psi = 0$, where $\hat{Z}$ is some differential operator, then, so long as $\hat{Z}$ is linear, if $\psi_A, \psi_B, \psi_C, …$ are solutions to the wave equation then so is their sum (or superposition), $\psi_A + \psi_B + \psi_C + \cdots$. So the addition rule makes sense from that perspective.

The new features of quantum mechanics, and indeed the larger part of the characteristically "quantum" behaviour of quantum systems, originate from the superposition principle, i.e., the addition rule, [2], which applies when there are alternative routes (paths) to a given outcome. The square modulus of [2] does not give the classical counterpart which would be $P_X = P_A + P_B + P_C + \cdots$. Instead, the square modulus of $\psi_A + \psi_B + \psi_C + \cdots$ also includes cross-terms like $\psi_A{}^* \psi_B$, etc. This is far more profound than a slight technical complication. In the classical expression, $P_A + P_B + P_C, + \cdots$ each term lies in the interval $[0,1]$ and hence each term increases the total. There is no negative probability. But in the expression $\psi_A + \psi_B + \psi_C + \cdots$ the individual wavefunctions have a phase factor: considered as vectors in the Argand plane they can be oriented in any direction. This means they can destructively interfere, so the combined probability might be zero, even though none of the contributing probability amplitudes is zero. In fact, as we shall see, precisely that is a common occurrence.

The potential for cancellation, or reinforcement, depending upon the relative phase of the terms in $\psi_A + \psi_B + \psi_C + \cdots$ is what gives rise to interference phenomena. These come about from superposition (i.e., the addition rule) together with the fact that wavefunctions are complex, and hence have a phase factor.

The counterintuitive behaviour of quantum interferometers is due to the alternative events A, B, C,... not being exactly mutually exclusive – though they may appear to be when the apparatus is interpreted classically. If the alternative paths through an interferometer were truly mutually exclusive – in other words if the photon or electron travelled along one path only to the exclusion of the others – then no interference would occur. Interference arises because the paths are not strictly mutually exclusive. Instead the photon or electron appears to travel along several different paths at once. This is why $\psi_A + \psi_B + \psi_C, + \cdots$ is needed to analyse the behaviour rather than $P_A + P_B + P_C, + \cdots$. This is explained more fully in my other book, *The Unweirding*.

I shall give some examples of the implications of the composition of probability amplitudes in §3.5, §3.8, §3.11 and §3.12, which will show how fundamental physical laws may be derived from the concept. But first I need to show how the wavefunction's phase factor, $e^{i\theta}$, typically varies in space and time.

3.4 The Origin of "Matter Waves"

I will present a rapid guide to the formulation of quantum mechanics in chapter 4. For now I want only to rationalise how the spatial dependence of the phase angle, θ, was arrived at by following the history. We have already been reminded of many key elements of the story in §3.2 as they appeared in Born's 1954 Nobel speech, but let us gather them together.

The story is usually started in 1900 with Planck's derivation of the blackbody radiation spectrum. However, it was Boltzmann, over twenty years earlier, who had first suggested that the energy levels of physical systems might be discrete rather than continuous. Planck deployed this idea, specifically hypothesising that electromagnetic radiation was emitted and absorbed in discrete 'packets' of energy proportional to its frequency, $E = hv$, where h is Planck's constant (its first appearance in physics).

It is remarkable that such a simple adjustment to one's thinking immediately solved what had been an intractable problem: assuming a continuous spectrum, blackbody radiation was predicted to be infinite. In contrast, Planck's quantisation condition led to results in excellent conformance with experiment.

However, it was not Planck but Einstein who took the further conceptual step of suggesting that light (or other electromagnetic radiation) of frequency v could *only* occur in quanta of energy $E = hv$. In 1905 Einstein applied this idea to the photoelectric effect, the work for which he was ultimately awarded the Nobel prize. (In the same year Einstein published his theory of relativity and his work on Brownian motion which clinched the existence of atoms). The idea explained why there was no photoelectric emission of electrons below a critical frequency, irrespective of the intensity of the light. It was not the total incident energy that mattered, but the energy in individual quanta (which were dubbed "photons" only 21 years later).

It was also Einstein who argued that, if these "photons" have energy $E = hv$ then they should also carry momentum, namely $p = hv/c$, and this idea had been given credence via its rationalisation of the Compton effect (the scattering of light by charged particles such as electrons).

1913 saw Bohr publish his model of the spectra of hydrogen which postulated that the electron in a stable atom could exist only in discrete energy levels. But Bohr's quantitative model started from the assumption that angular momentum was quantised in integer multiples of $h/2\pi$ (the quantity now denoted $\hbar$).

It was de Broglie in 1923 who made the major conceptual break-through (as part of his PhD work) by postulating that all matter could be regarded as having wave-like properties. Thereafter things happened with prodigious speed over the next 5 years.

In 1925, Born, Heisenberg and Jordan at Göttingen developed matrix mechanics. In 1926 Schrodinger published his famous wave equation. In 1927

Dirac unified the various approaches to quantum mechanics and in 1928 produced his relativistic equation. 1927 also saw the experimental observation of electron diffraction, as predicted by de Broglie and (by then) Schrodinger's and Dirac's wave theories. In the same year, Heisenberg published his famous Uncertainty Principle.

The advantage of the wave equation approach of Schrodinger is that matter-waves were explicit – and the wavefunction was explicitly and essentially complex-valued. And, as we have seen, Born provided the physical interpretation of Schrodinger's wavefunction, namely that $|\psi|^2$ is the particle's probability density in space.

So, the wavefunction, ψ, is a (sort of) square root of probability…with a phase factor.

The developments described above allowed physicists to identify how the phase angle, θ, varied in space, and this was especially clear in the wave equation formulations. For variety, though, let's derive this dependence from the matrix mechanics formulation without using a wave equation. We have seen in §3.2 that Born first published, in 1925, what we now call the canonical commutation relation,

$$\hat{p}\hat{q} - \hat{q}\hat{p} = \frac{h}{2\pi i} = -i\hbar \tag{3.4.1}$$

The matrices representing momentum ($\hat{p}$) and position ($\hat{q}$) do not commute. Let's do as conventional quantum texts do and adopt a representation of these observables which act on functions. "Position", $\hat{q}$, just becomes the real variable x (confining attention to one dimension for now). The momentum operator becomes a differential operator with the commutation property, for "any" function $f(x)$,

$$(\hat{p}x - x\hat{p})f(x) = -i\hbar f(x)$$

We see that a suitable representation of $\hat{p}$ with this property is,

$$\hat{p} = -i\hbar \frac{\partial}{\partial x} \tag{3.4.2}$$

I now anticipate a key postulate of quantum mechanics, to be presented more fully in chapter 4, that the possible values that an observable can take upon measurement are the eigenvalues of the corresponding operator. In this case this means that the possible values of momentum, p, are such that there are particular wavefunctions $\psi(p, x)$ such that,

$$\hat{p}_x \psi(p_x, x) = -i\hbar \frac{\partial}{\partial x} \psi(p_x, x) = p_x \psi(p_x, x)$$

where p_x is a number, the x-momentum. Now this can be integrated directly, thus,

$$-i\hbar \int \frac{d\psi}{\psi} = -i\hbar \cdot log(\psi) = p_x \int dx = p_x x + A$$

where A is some constant. This gives,

$$\psi = Be^{ip_x x/\hbar} \qquad (3.4.3)$$

where B is some other constant. Actually, because we were dealing with the partial derivative in (3.4.2), B is only constant wrt x. It may be a function of y, z, t. By starting with the equivalent of (3.4.2) in the other spatial variables, and noting that the momentum operators commute with each other, we arrive at how the wavefunction varies in 3D space, assuming its momentum is well defined in all three coordinate directions,

$$\psi = Ce^{i(p_x x + p_y y + p_z z)/\hbar} \qquad (3.4.4)$$

The equivalent commutation relation to (3.4.1) but between energy and time matrices/operators, is,

$$\hat{E}\hat{t} - \hat{t}\hat{E} = i\hbar \qquad (3.4.5)$$

Which results in a differential representation of the energy operator as,

$$\hat{E} = i\hbar \frac{\partial}{\partial t} \qquad (3.4.6)$$

And so, in the same way, assuming the energy is well defined as well as all three momenta, the wavefunction – now dependent on time as well as position – becomes,

$$\psi = De^{i(p_x x + p_y y + p_z z - Et)/\hbar} \qquad (3.4.7)$$

where D is now a constant independent of x, y, z, t. The phase angle can more compactly be written $k \cdot x = k^\mu x_\mu$ where we have adopted 4-vector notation and the Minkowski metric, with $x_\mu = (t, \vec{r})$ and the wavevector is $k^\mu = (\omega, -\vec{k})$, where the angular frequency is given by $\omega = E/\hbar$ and the wave 3-vector by $\vec{k} = \vec{p}/\hbar$. (Don't panic, see §3.7 for a rapid review of relativistic notation).

So, finally, the all-important phase angle which we introduced in (3.3.1) simply by taking a (sort of) square root has now been pinned down to be,

$$\theta = k \cdot x = \vec{k} \cdot \vec{r} - \omega t \qquad (3.4.8)$$

It is this spatial and temporal dependence which produces the characteristically wave-like properties, noting that the imaginary exponential is just a sum of cosine and sine terms by de Moivre's theorem (see 2.14.3) and hence is oscillatory, these functions being periodic.

In the interests of completeness, and to avoid misleading the reader, please note that a "plane wave" dependence of the wavefunction on space and time, i.e., (3.4.4), is valid only for free particles. The wavefunction for particles in a bound state, e.g., atomic electrons, is very different (rather obviously as they are confined to the atom, whereas the magnitude of (3.4.7) does not diminish at large distances).

The rest of this chapter illustrates some important consequences of wavefunctions which have the phase factor $e^{ik \cdot x}$.

3.5 The Double Slit Experiment

Figure 3.1 Double Slit Experiment

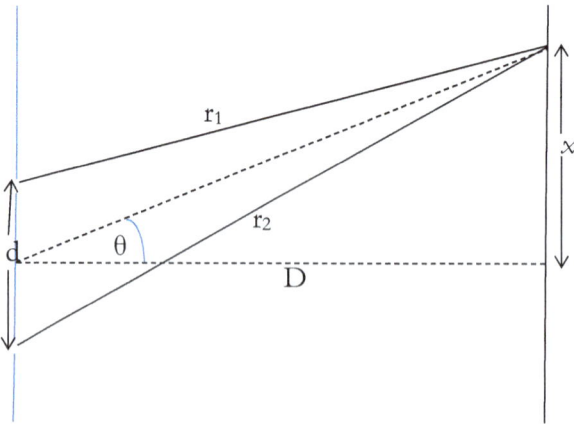

I suspect that one would be excommunicated from the society of physicists if one were to write a book on quantum mechanics and not include the iconic, and instructive, double slit experiment. It is the equivalent with matter particles of Young's double-slit interference experiment with light, which did much to confirm the wave nature of light early in the nineteenth century. The practical details need not concern us. Carrying out the experiment using electrons and real slits is far from easy and it was many decades after the formulation of basic quantum mechanics by de Broglie, Schrodinger and the rest before an experiment was carried out in this manner. (However, the first experimental demonstration of the wave-like properties of electrons was achieved, as early as 1927, by diffraction from crystals, in which the planes of atoms function as the diffraction grating).

The essence of the experimental arrangement is that a monochromatic beam of electrons (i.e., a beam in which all the electrons have the same energy, and hence, crucially, the same wavelength) is incident upon a screen with two very narrow slits. Diffraction within each slit effectively turns each slit into a point source. The most impressive experiments are performed with incident beams of such low intensity that only one electron is within the apparatus at any time. Consequently, the interference observed is the wavefunction of a single electron interfering with itself (as opposed from the wavefunctions of two electrons interfering with each other). The interference formula is familiar from school physics and is easily derived in terms of the wavelength, λ, as follows.

Familiarity makes it easy to overlook that this is an application of the idea of probability amplitudes and how they combine (interfere) according to the rules of §3.3.1. Relative to the phase at the slits, the phase angle of the top beam after travelling distance r_1 is kr_1 where the wavenumber, k, is $2\pi/\lambda$. Similarly, the phase angle of the lower beam after travelling distance r_2 is kr_2 as the beams have the same wavelength (i.e., the same energy). We now appeal to combination rule [2] of §3.3.1. The combined wavefunction at point x on the screen is thus proportional to the sum of the two probability amplitudes (because an electron may go by the upper path OR by the lower path, and OR = addition),

$$\psi(x) \propto e^{ikr_1} + e^{ikr_2}$$

(I'm ignoring the overall magnitude: experts can insert a factor of $1/\sqrt{2}$ if they wish). The slits are a distance d apart. The interference pattern appears on the screen to the right, and we focus on the point a distance x from the mid-point and hence at angle θ, as shown. Bright lines occur where the beams from the two slits are in-phase, i.e., when the path lengths r_1 and r_2 differ by a whole number of wavelengths, i.e., $|r_2 - r_1| = n\lambda$ for some integer n. Dark bands occur in-between, where the two path lengths differ by $(n + 0.5)\lambda$. Pythagoras gives $r_2^2 = D^2 + (x + d/2)^2$ and $r_1^2 = D^2 +$

$(x - d/2)^2$ and hence $r_2^2 - r_1^2 = 2xd$. Using $2x/(r_2 + r_1) \approx \sin \theta$ gives the condition for bright bands to be $n\lambda = \pm d \sin \theta$. <u>QED</u>.

3.6 Feynman Path Integrals

The double slit experiment is the simplest example of a situation in which the particle may travel by two paths and the physical result depends on the superposition (i.e., the addition) of the two probability amplitudes. The Mach-Zehnder interferometer is another simple case where there are two contributing paths. (See my other book, *The Unweirding*, for many examples of how Mach-Zehnder interferometers illustrate the puzzling behaviour of pure quantum states). This simplicity, i.e., that only two paths are involved, is the attraction of these arrangements for experiments which aim to illustrate and clarify quantum behaviour.

Figure 3.2: Feynman's Path Integral

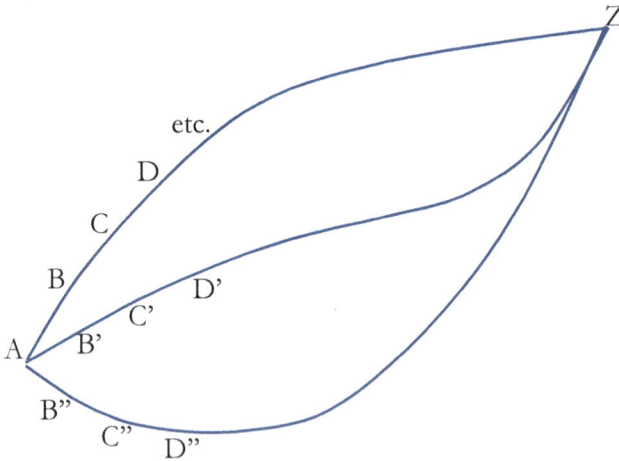

More generally, a particle may move from A to Z via many paths. In free space there is an indenumerable infinity of such. (Readers may amuse themselves by considering the associated cardinality, as part of which it will be necessary to refine what counts as a "path" – but we will not need such sophistication). Richard Feynman's famous 1948 spacetime approach to quantum mechanics took seriously the idea that a particle might move from A to Z via any continuous path, however indirect. In this approach to basic physics one ignores Newton's laws of motion and any other formulation of dynamics. Instead, each path is assigned a probability amplitude calculated using the product rule from the sequence of probability amplitudes associated with every small increment along the path in question. The overall probability amplitude to arrive at Z starting from A is obtained by using the addition rule, i.e., by summing the probability amplitudes for every alternative path.

Referring to Figure 3.2, the probability amplitude of the upper path is $\psi_{AB}\psi_{BC}\psi_{CD}$ ... where each factor is the probability amplitude for a small increment along the path, starting with AB, then BC, etc. The probability amplitude for the other paths is found in similar fashion. The overall probability amplitude to arrive at Z starting from A is then,

$$(3.6.1)$$

$$\psi_{AZ} = \psi_{AB}\psi_{BC}\psi_{CD} \ldots + \psi_{AB'}\psi_{B'C'}\psi_{C'D'} + \psi_{AB''}\psi_{B''C''}\psi_{C''D''} + \ldots$$

where the sum extends over "all" possible paths. (I draw a veil over exactly what "all" means, but note that continuity is assumed but differentiability is certainly ***not*** assumed. Gradients may be highly discontinuous).

Feynman was the first to realise (following an idea of Dirac's) that the phase angle of the probability amplitude could be identified with the classical action (divide by $\hbar$). His path integral, in a non-relativistic formulation, then showed how classical dynamics (Newton's laws) results from the most probable path. The most probable path becomes so overwhelmingly more probable than any others in the classical limit that it becomes a unique, deterministic path for all practical purposes. Moreover, the same analysis also allowed Feynman to derive the Schrodinger equation, thus obtaining the classical behaviour and the quantum behaviour from the same analysis, the former being the limiting case of the latter.

The path integral approach was later adopted in relativistic quantum field theory (QFT) and became its standard technique. I shall avoid action integrals, Lagrangians and field theory here. However, the path integral, (3.6.1), will be very enlightening without such embellishments.

3.7 A Spot of Relativity Revision

Derivation of the implications of the path integral formulation of the wavefunction, (3.6.1), is best done in the relativistic context. Here I reprise the basics of (special) relativity. The starting point is the Lorentz transformation which connects the space-time coordinates measured by two observers in different states of (uniform) relative motion. Thus, if observer S sees observer S' moving at velocity u in the x direction then,

$$x = \gamma(x' + ut') \tag{3.7.1a}$$

$$t = \gamma(t' + ux') \tag{3.7.1b}$$

Coordinates perpendicular to the velocity are the same for the two observers, $y = y'$, $z = z'$. Here (and throughout) I have set the speed of light to be unity ($c = 1$), so speeds of observers or of particles with non-zero rest mass lie in the range $(-1, +1)$, where the round brackets mean "not inclusive". The famous relativistic gamma factor is,

$$\gamma = \frac{1}{\sqrt{(1-u^2)}} \tag{3.7.2}$$

The gamma factor has a minimum of 1 at speeds slow compared with light speed ($|u| \ll 1$) and is divergent as $u \to \pm 1$.

The reader should confirm that the "Minkowski metric" is invariant, i.e., that $t'^2 - x'^2 = t^2 - x^2$ and also $t'^2 - r'^2 = t^2 - r^2$ where r is the 3D distance. The latter form is true irrespective of the direction of motion in 3D space.

If the observer S' is comoving with a particle, i.e., the particle is also moving at uniform velocity u in the x direction, and the coordinates above relate to the particle's location, we may take the coordinate of the particle wrt S' to be $x' = 0$. The time coordinate t' is then the time experienced by the particle itself, i.e., in its own rest frame. This time is known as the "proper time", denoted τ. From (3.7.1a,b) we see that the coordinates observed by S are, in terms of the proper time, $t = \gamma\tau$ and $x = \gamma u\tau = ut$.

The first of these expresses time dilation, i.e., at sufficiently high speeds, time intervals experienced by the particle itself are smaller than the time interval as it appears to observer S who sees the particle as moving, i.e., $\tau = \frac{t}{\gamma} < t$. This phenomenon (mind boggling though it appeared in 1905) is routinely observed in the decay of unstable particles whose lifetimes as measured in the laboratory frame increase with speed precisely as predicted by the simple formula $t = \gamma\tau$, the true lifetime in their rest frame being measured by τ.

In the limit, particles moving at light speed experience no passage of time. (Consequently, zero rest mass particles cannot be unstable).

Energy and momentum transform between the observers in the same way as the coordinates, thus,

$$p_x = \gamma(p_x' + uE') \tag{3.7.3a}$$

$$E = \gamma(E' + up_x') \tag{3.7.3b}$$

The perpendicular components of momentum are the same as seen by the two observers, $p'_y = p_y, p'_z = p_z$. In the same way as the Minkowski metric, it follows that $E^2 - p_x^2$ is invariant, i.e., $E'^2 - p'^2_x = E^2 - p_x^2$. In 3D, this becomes $E'^2 - p'^2 = E^2 - p^2$ where $p = |\vec{p}|$ irrespective of the direction of relative motion. The combination $(E, \vec{p})$ is called the momentum 4-vector, or simply the 4-momentum.

Similarly, $(t, \vec{r})$ is the 4-vector of "position" in spacetime. To distinguish a point in space, defined by a given $\vec{r}$, a "point" in 4D spacetime, defined by a given $(t, \vec{r})$, is referred to as an "event", because it specifies both a spatial point and a time. 4-vectors have an inner product (scalar product or dot product) defined by, for example, $(t, \vec{r}) \cdot (E, \vec{p}) \equiv Et - \vec{p} \cdot \vec{r}$. Consequently the invariance of $E^2 - p^2$ and of $t^2 - r^2$ under Lorentz transformations (including rotations) justifies the term "scalar product" as these quantities are equal to $(E, \vec{p}) \cdot (E, \vec{p})$ and $(t, \vec{r}) \cdot (t, \vec{r})$ respectively.

3.8 Particles Have a Fixed Mass

That $E'^2 - p'^2 = E^2 - p^2$ is invariant between observers in relative uniform motion means that this quantity is found to be the same whichever inertial observer measures it. Moreover, the result they get will be E_{rf}^2, where E_{rf} is the energy measured in the rest frame (because, by definition, the momentum is zero in that frame). But this only shows that, for any given instantiation of a particle, all observers will agree on this quantity. It does not prove that every instantiation of the particle will display the same value for E_{rf}. That is, it does not prove that E_{rf} is always the same for every instance of a given type of particle and hence can be interpreted as a particle property (specifically, as we shall see, the particle's mass).

Indeed, in fundamental particle physics it is usually just assumed that a given particle type has a specific, well-defined, constant mass. (Experts should note that the mere identification of mass with the eigenvalue of the Casimir operator of the Poincaré group does not show that every instance of a particle has the same mass, nor even that the fundamental particles have a discrete mass spectrum). Unstable particles do not have a well-defined mass so much as a distribution of mass, the width of the distribution being inversely related to the particle's half-life. But even for unstable particles, the mean mass is a well-defined constant. It is true that in gauge field theories, specifically in the standard model of particle physics, the masses of fundamental particles may not appear by *fiat* but arise through coupling to the Higgs field. But the resulting particle masses are in one-to-one correspondence with the Yukawa coupling constants, so the buck is merely passed from an assumed mass constant to an assumed coupling constant.

The fundamental question is: why is the particle mass spectrum discrete rather than continuous? Why does each distinct fundamental particle, distinguished by various other quantum numbers, magnetic moments, etc., have a well-defined and constant value for E_{rf}, which we can then denote the mass, m? I am not aware that there is generally perceived to be a need to justify the assumption, though I think there is such a need. There is a very simple argument which we are now in a position to present.

Consider a particle propagating from position $\vec{r}_1$ at time t_1 to position $\vec{r}_2$ at time t_2. This defines an average velocity, $\vec{u} = (\vec{r}_2 - \vec{r}_1)/(t_2 - t_1)$, and hence a relativistic γ factor], (3.7.2). The energy and momentum seen from the "lab" frame, i.e., the frame in which the particle moves with velocity $\vec{u}$ is $(E, \vec{p}) = E_{rf}(\gamma, \gamma\vec{u})$. So the energy and momentum are not specified by the conditions of the problem, i.e., that the particle moves from position $\vec{r}_1$ at time t_1 to position $\vec{r}_2$ at time t_2. There is a further degree of freedom, namely the value of E_{rf}.

Let us deploy *Reductio ad Absurdum* and assume that E_{rf} is not a fixed constant, i.e., the rest mass, but is instead arbitrary. We now argue that if this were the case, then the particle could propagate between the two specified spacetime events via a range of energies and momenta according to different chosen values of E_{rf}.

Now the probability amplitude associated with any assumed energy and momentum is proportional to $e^{ik \cdot x} = e^{i(E,\vec{p}) \cdot (t,\vec{r})} = e^{i(Et - \vec{p}.\vec{r})}$. But we can simplify this by using comoving (i.e., rest frame) coordinates in terms of which we have $Et - \vec{p}.\vec{r} = E_{rf}\tau$. So, the probability amplitude for a given E_{rf} is proportional to $e^{i\tau E_{rf}}$.

But since, by assumption, we may choose "any" value for E_{rf}, we need to use combination rule 2 and integrate the probability amplitude over E_{rf}. Hence,

$$\text{Probability amplitude} = \int e^{iE_{rf}\tau}dE_{rf} \tag{3.8.1}$$

I have deliberately left the range of integration vague but it is clear that if the range is over an integral multiple of $2\pi/\tau$ then the integral will be zero. This will only fail to be the case if $\tau = 0$. The same conclusion will be shown to be valid in §3.10 even if the range of integration is $[-\infty, +\infty]$, again with the exception of $\tau = 0$.

So, we have shown that if E_{rf} is a degree of freedom the particle has zero probability of propagating, unless possibly when $\tau = 0$. Consequently, to avoid the paradoxical conclusion that a particle cannot move, and provided that $\tau \neq 0$, it follows that a particle must have a constant, fixed value for E_{rf} which is the rest mass, m. QED.

To paraphrase, the argument implies that the particle mass spectrum is discrete, and a given particle type has a single, well-defined mass.

This implies that, because $E^2 - p^2$ is invariant and necessarily equals $E_{rf}^2 = m^2$ in the rest frame where momentum is zero, it follows that in any inertial reference frame $m = \sqrt{E^2 - p^2}$.

The exceptional case $\tau = 0$ relates to zero rest mass particles which always have zero elapsed proper time (they travel on "null rays" in the jargon).

3.9 The Significance of Fixed Mass

The significance of a given particle type having a definite, constant mass is that, because $m = \sqrt{E^2 - p^2}$ in any inertial reference frame, the energy of a particle is known if the mass and the magnitude of momentum are known. Specifically, assuming positive energy, $E = \sqrt{m^2 + p^2}$. This is the relativistic expression for total energy (which, of course, varies according to the observer's state of motion). Kinetic energy is the amount by which the total energy exceeds the rest mass, $KE = \sqrt{m^2 + p^2} - m$. For speeds small compared to light speed we have $p \ll m$ and the kinetic energy can be approximated using the binomial theorem.

$$KE = \sqrt{m^2 + p^2} - m \approx m\left(1 + \frac{1}{2}\frac{p^2}{m^2}\right) - m = \frac{p^2}{2m} = \frac{1}{2}mv^2$$

which is the familiar non-relativistic kinetic energy.

3.10 Introducing the Dirac Delta Function

A useful device, introduced by Dirac, is the delta function. It is not a function in the usual sense but a generalised function - see Lighthill's book for details of the rigorous mathematics. Here I am content with the usual heuristic treatment in physics texts. The most common representation of the Dirac delta function is this,

$$\delta(x - x_0) = \frac{1}{2\pi}\int_{-\infty}^{+\infty} e^{ik(x-x_0)}\, dk \tag{3.10.1}$$

It is no ordinary function because it is zero for all real values of x except exactly at $x = x_0$ where it is infinite. The usefulness of this "function" is the following property, which holds for "any" other function $g(x)$,

For any $a < x < b$: $\int_a^b g(x')\delta(x' - x)\,dx' = g(x)$ (3.10.2)

If x lies outside the range $[a, b]$ then the integral is zero.

These observations confirm that the integral (3.8.1) over infinite limits is indeed zero if $\tau \neq 0$.

For the faint hearted (or, more properly, the rightly cautious) the Dirac delta function can be considered as the limit of a Gaussian, or normal distribution, as its width is reduced to zero, thus,

$$\delta(x - x_0) = LIM_{\sigma \to 0} \left\{ \frac{1}{\sigma\sqrt{2\pi}} \int_{-\infty}^{+\infty} e^{-\frac{(x-x_0)^2}{2\sigma^2}}\,dx \right\}$$

There are many other representations of the Dirac delta function, but these will suffice. The derivative of the delta function, $\delta'(x - x_0)$, has the interesting property that it pulls out the derivative of an integrand,

For any $a < x_0 < b$: $\int_a^b g(x)\delta'(x - x_0)\,dx = -\frac{dg(x_0)}{dx}$ (3.10.3)

This may be shown using integration by parts, or simply by noting that, by definition of the derivative, the LHS can be written,

$$LIM_{\varepsilon \to 0} \int_a^b g(x)\,\frac{\delta(x+\varepsilon-x_0)-\delta(x-x_0)}{\varepsilon}\,dx = \frac{g(x_0-\varepsilon)-g(x_0)}{\varepsilon} = -\frac{dg(x_0)}{dx}$$

3.11 Newton's First Law

Newton's First Law states that any free body, not acted upon by a force, maintains constant velocity (constant speed in a straight line). This includes being stationary as a special case. For a non-zero rest mass body we can now show that this arises from the path integral over probability amplitudes. Suppose, as we did in §3.8, that a particle starts from position $\vec{r}_1$ at time t_1 and ends at position $\vec{r}_2$ at time t_2. But now we consider what would result from the path integral approach if the particle did not move in a straight line with constant speed (*Reductio ad Absurdum* again). The paths that it might take are legion, but it suffices to consider a path which consists of two straight line segments, as shown in Figure 3.3.

Figure 3.3: Deriving Newton's First Law

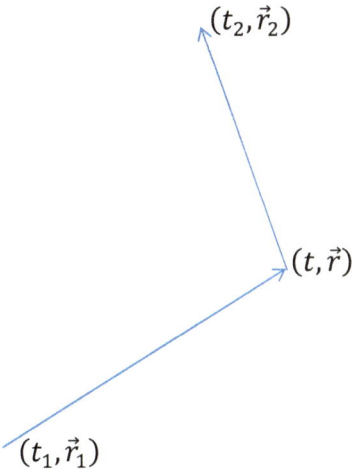

The probability amplitude for the first segment, from point 1 to the intermediate point, is proportional to $e^{im\tau_1}$ where,

$$\tau_1 = \sqrt{(t - t_1)^2 - |\vec{r} - \vec{r_1}|^2} \tag{3.11.1}$$

The probability amplitude for the second segment, from the intermediate point to point 2, is proportional to $e^{im\tau_2}$ where,

$$\tau_2 = \sqrt{(t_2 - t)^2 - |\vec{r_2} - \vec{r}|^2} \tag{3.11.2}$$

The probability amplitude for starting at $\vec{r_1}$ at time t_1 and ending at $\vec{r_2}$ at time t_2 and propagating via two straight paths joined at event $(t, \vec{r})$ is proportional to the product of the probability amplitudes for each segment, i.e., $e^{im(\tau_1 + \tau_2)}$. The overall probability amplitude of propagating between these fixed start and end points by two straight lines via any intermediate point is therefore the integral of $e^{im(\tau_1 + \tau_2)}$ over all $(t, \vec{r})$, i.e.,

$$\text{Probability amplitude} = \int e^{im(\tau_1 + \tau_2)} \, dt dx dy dz$$

For almost all paths, small variations of the intermediate point $(t, \vec{r})$ will cause the exponent in the above integrand to vary rapidly, and hence result in strong cancellations. This will only fail to be the case where the exponent is stationary (i.e., has a turning point). The so-called method of stationary phase thus indicates that the only paths which will have a net contribution to the overall probability amplitude will be those for which the exponent is locally stationary.

Note that the spatial integral is over the volume, i.e., over all three spatial directions of $\vec{r}$. This integral is very different from those of (3.8.1) and (3.10.1) in which the exponent was a linear function of the integration variable. Here the exponent, $im(\tau_1 + \tau_2)$, is a non-linear function of the integration variables. In particular the exponent has a non-zero minimum value because the total proper time $\tau_1 + \tau_2$ cannot be zero because the path must connect two events and the particle is assumed to have non-zero mass, $m \neq 0$ (i.e., a non-zero rest mass particle cannot travel from one place to another in zero time, even as seen in its own rest frame). This minimum in the exponent occurs where the derivatives of $\tau_1 + \tau_2$ wrt all four variables, t, x, y, z, are zero. Consider the x derivatives which follow from (3.11.1-2),

$$\frac{\partial \tau_1}{\partial x} = \frac{(x_1 - x)}{\tau_1} \quad \text{and} \quad \frac{\partial \tau_2}{\partial x} = \frac{(x_2 - x)}{\tau_2}$$

Hence,
$$\frac{\partial(\tau_1 + \tau_2)}{\partial x} = \frac{(x_1 - x)}{\tau_1} + \frac{(x_2 - x)}{\tau_2} = 0$$

Hence,
$$\frac{(x - x_1)}{\tau_1} = \frac{(x_2 - x)}{\tau_2} \quad \text{or} \quad \frac{(x - x_1)}{(x_2 - x)} = \frac{\tau_1}{\tau_2}$$

By symmetry, setting the derivatives wrt y, z, t to zero gives us,

$$\frac{(x - x_1)}{(x_2 - x)} = \frac{(y - y_1)}{(y_2 - y)} = \frac{(z - z_1)}{(z_2 - z)} = \frac{(t - t_1)}{(t_2 - t)} = \frac{\tau_1}{\tau_2} \qquad (3.11.3)$$

This says that the ratio of the coordinate increases over the first and second segments is the same for all four coordinates. In other words, the trajectory which minimises the proper time, $\tau_1 + \tau_2$, and hence makes the exponent in the integral stationary, is a straight line with the same speed in the two segments. This trajectory is the only one which does not involve cancellations in the integral. Ergo, the particle travels in a straight line at constant speed, and the (classical) trajectory is not really kinked as suggested by Figure 3.3 but straight.

There is more to this than merely the observation that the shortest distance between two points is via a straight line. That is indeed pertinent to the result – but the path integral over probability amplitudes rationalises why it is the minimum distance (or, in this case, minimum proper time) which dominates.

3.12 The Conservation of Energy and Momentum

The conservation of momentum, and hence the conservation of energy, for a set of particles which are not interacting (with themselves or an external agency) follows trivially from the preceding observations that each particle propagates with constant velocity and constant mass, hence constant energy and momentum. But what if the particles interact? To avoid action at a distance, particles may be assumed to interact only when they meet at the same spacetime event. In classical physics, the interaction between (say) two charged particles was considered as mediated by an electromagnetic field. One charge was considered to create an electromagnetic field throughout space, and the other particle responded according to the field acting at its location (and vice-versa). In the quantum version of this, the fields are considered as composed of field quanta (photons) and the interaction between two charged particles is mediated by these quanta. There is therefore a "quantum" of interaction, which consists of just one photon interacting at one point with one charged particle. (A classical electromagnetic field can be considered as the superposition of a divergently large number of individual multi-photon states).

Figure 3.4 depicts (say) an incoming electron with 4-momentum $p_1 = (E_1, \vec{p}_1)$ which interacts with a photon at spacetime event $x = (t, \vec{r})$ so that the electron emerges with 4-momentum $p_2 = (E_2, \vec{p}_2)$ and the photon with 4-momentum $q = (\omega, \vec{q})$.

Figure 3.4: The Conservation of Energy and Momentum

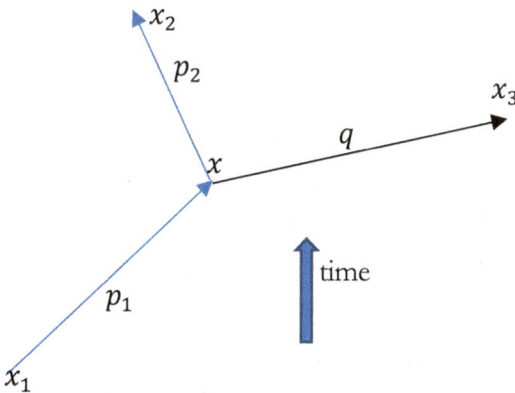

We consider fixed spacetime events x_1, x_2, x_3 from which the particles start or finish, as illustrated. The probability amplitude for the incoming electron to propagate from x_1 to the intermediate event at x is proportional to $e^{ip_1\cdot(x-x_1)}$, where the exponent is the scalar product of the two 4-vectors and so is shorthand for $i[E_1(t - t_1) - \vec{p}_1 \cdot (\vec{r} - \vec{r}_1)]$. The order of the two events is defined by the temporal order. The probability amplitude for the emerging electron to propagate from the interaction event to x_2 is similarly $e^{ip_2 \cdot (x_2 - x)}$, and the corresponding probability amplitude for the emerging photon is $e^{iq \cdot (x_3 - x)}$. The overall probability amplitude for the interaction sequence is therefore proportional to,

Probability amplitude $\propto e^{ip_1 \cdot (x - x_1)} e^{ip_2 \cdot (x_2 - x)} e^{iq \cdot (x_3 - x)}$

The intermediate event may take place "anywhere" in spacetime, so we need to use Rule 2 (superposition) to find the final probability amplitude for the interaction by integrating over the event location, x (which involves four degrees of freedom), thus,

$$\text{Probability amplitude} = e^{i(-p_1 \cdot x_1 + p_2 \cdot x_2 + q \cdot x_3)} \int e^{ix \cdot (p_1 - p_2 - q)} d^4 x \qquad (3.12.1)$$

This probability amplitude integrates to zero whenever $(p_1 - p_2 - q)$ is not zero. Thus, there can be a non-zero probability for this interacting taking place only if $(p_1 - p_2 - q) = 0$, i.e., if the incoming energy and momentum vector equal those emerging afterwards, $p_1 = p_2 + q$. In the language of §3.10, the integral is a Dirac delta function and the probability amplitude is proportional to $\delta^4(p_1 - p_2 - q)$, where this notation stands for the product of four delta functions, one for each component of its argument's 4-vector. Specifically,

$$\delta^4(p_1 - p_2 - q) = \delta(E_1 - E_2 - \omega)\delta(p_{1x} - p_{2x} - q_x)\delta(p_{1y} - p_{2y} - q_y)\delta(p_{1z} - p_{2z} - q_z)$$

Figure 3.4 is essentially a Feynman diagram. Although we have argued from the perspective of the path integral over probability amplitudes, the same conclusion for the conservation of 4-momentum at "vertex interactions" is arrived at in relativistic quantum field theory (QFT). The same integral as in (3.12.1), and hence the result $\delta^4(p_1 - p_2 - q)$, is found via the machinery of QFT, specifically in the vacuum expectation value of products of quantum fields.

What I have not rationalised here is that this "vertex interaction" involves a continuous electron line (the incoming line becoming the emerging line) whereas the photon is actually created at the event. The machinery of QFT does this, but we'll not dwell on the matter here (beyond noting that it arises from the particular product of quantum fields which represents the electromagnetic coupling). The number of lines meeting at the event would not affect the conclusion that the total incoming 4-momentum must equate with the total outgoing 4-momentum, however many particles contributed to either.

It is immediately clear that one may now have multiple interactions with many "vertices", each conserving 4-momentum, so that the net effect is the that total incoming 4-momentum and the total outgoing 4-momentum will be equal by virtue of this being the case at every vertex. Hence, the conservation of energy and momentum apply to real physical events at all scales.

3.13 Newton's Second and Third Laws

Newton's Second and Third Laws of Motion are couched in terms of forces, but force is not a concept that occurs in quantum mechanics. However, the two laws together have an immediate consequence expressible without the use of forces. The Second Law says that the net (vectorial) force on a body equates with the rate of change of its (vectorial) momentum. The Third Law says that when two bodies interact, the force due to the interaction on one is equal and opposite to the force on the other. It follows that when two bodies interact, the rate of change of momentum on one is equal and opposite to that of the other, i.e., that their combined momentum does not change. Hence, the deduction of the conservation of energy and momentum from the path integral approach incorporates, in this limited sense, the import of Newton's Second and Third Laws.

3.14 References

[1] Feynman, R.P. (1948), "*Space-Time Approach to Non-Relativistic Quantum Mechanics*", Reviews of Modern Physics **20**(2), 267

[2] Lighthill, M.J. (1959), "*Introduction to Fourier Analysis and Generalised Functions*", Cambridge University Press.

[3] Pauli, W. (1926). *Über das Wasserstoffspektrum vom Standpunkt der neuen Quantenmechanik*. Zeitschrift für Physik A Hadrons and nuclei volume 36, pages336–363 (1926).

[4] Schrödinger, E. (1926). *Quantisierung als Eigenwertproblem*. Annalen der Physik. Leipzig 79 (1926) 361

4

The Square Root of Not

Here we take a brief look at the theoretical basis of quantum computation, but only for the purpose of explaining how the operation of logical negation ("NOT") can have a square root. That will lead naturally to a discussion of non-bivariate logic, the Law of the Excluded Middle and the challenge its abandonment would pose to mathematical proof. I draw a distinction between the usefulness of the inherently non-bivariate nature of quantum states themselves and the observable "truth values" derived from them. It is the latter which drives their utility in human reasoning. It is appropriate to raise some related logical and philosophical issues, and I do so to illustrate how the power of mathematics and physics is crucially related to the deliberate limiting of their sphere of discourse. Thus, the power of mathematics and physical science fails if an attempt is made to understand issues beyond their limited range of applicability, such as transcendent or metaphysical matters.

We shall not venture very far into the issues raised by quantum computers. But it is necessary to set up the basic formulation of quantum mechanics to discuss their essential features. In particular, it is the phenomenon of superposition which is key – the same phenomenon which is primarily responsible for imbuing quantum mechanics with its celebrated weirdness.

For people interested in a full exposition of how quantum mechanics is formulated, please see chapter 2 of my other book, *The Unweirding*. For our present purposes a very brief account will suffice, and for simplicity I will confine attention to cases where measurements have a discrete set of possible outcomes (often characteristic of quantum mechanics).

To avoid misunderstanding, I make the point immediately that what we shall be considering in this chapter is **not** so-called "quantum logic". As an alternative to traditional methods of propositional inference, quantum logic has been largely, if not entirely, discredited. Quantum computation, on the other hand, is something entirely different and very much a live field of endeavour.

4.1 Basic Formulation of Quantum Mechanics

Suppose the possible outcomes of measuring an observable Q are the real numbers $q_1, q_2, q_3, \dots$. In quantum mechanics the most general **pure** quantum state of such a physical system can be represented mathematically by,

$$|\psi\rangle = a_1|q_1\rangle + a_2|q_2\rangle + a_3|q_3\rangle + \cdots \tag{4.1.1}$$

This is Dirac notation for Hilbert space. Again I refer you to chapter 2 of *The Unweirding* for a fuller discussion. Here's a rapid explanation. A Hilbert space is a type of vector space equipped with an inner product. Don't panic, it's very simple…

4.1.1 Vector Space

Firstly, let's define a vector space. Mathematicians will forgive me if I define only vector spaces over the complex numbers. A vector space is a set of objects called vectors, which we write in the notation $|\psi\rangle, |\phi\rangle$, etc., such that any vector can be multiplied by a complex number, c, to give another vector within the vector space, $c|\psi\rangle$, and any two vectors can be "added" to give another vector within the space, $|\psi\rangle + |\phi\rangle$. So, if we denote the vector space by $\mathcal{H}$ then, in formal notation,

$$|\psi\rangle \in \mathcal{H} \Rightarrow c|\psi\rangle \in \mathcal{H} \quad \text{and} \quad |\psi\rangle, |\varphi\rangle \in \mathcal{H} \Rightarrow |\psi\rangle + |\varphi\rangle \in \mathcal{H}$$

(where $a \in S$ means "a is an element of the set S", and $\Rightarrow$ means "implies", in the mathematical sense of being logically necessary). In the jargon, the vector space is closed under the operation of addition and the operation of multiplication by a scalar (i.e., by a complex number).

Multiplication by a complex number and vector addition are also required to obey certain conditions such as associativity, commutativity and distributivity. So we have, for example,

Commutative:
$$|\psi\rangle + |\phi\rangle = |\phi\rangle + |\psi\rangle$$

Associative:
$$|\psi\rangle + (|\phi\rangle + |\omega\rangle) = (|\phi\rangle + |\psi\rangle) + |\omega\rangle$$

This associativity means that vector addition can be written without brackets, so an expression like (4.1.1), above, is unambiguous.

Distributive 1:
$$a(|\psi\rangle + |\phi\rangle) = a|\psi\rangle + a|\phi\rangle$$

Distributive 2:
$$(a + b)|\psi\rangle = a|\psi\rangle + b|\psi\rangle$$

Finally, there must also be a "zero vector", $|zero\rangle$, such that $|\psi\rangle + |zero\rangle = |\psi\rangle$ for all vectors, and it follows that $0|\psi\rangle = |zero\rangle$. (I write $|zero\rangle$ rather than $|0\rangle$ because the notation $|0\rangle$ is often reserved for a non-zero vector, e.g., a ground state. So, $0|\psi\rangle$ is not, in general, equal to $|0\rangle$).

Arbitrary vectors are conveniently written in terms of a set of vectors called a basis. If the vectors $|q_1\rangle, |q_2\rangle$ … are a basis, then complex coefficients $a_1, a_2,$ … always exist so that any vector can be written like (4.1.1). Moreover, the coefficients $a_1, a_2,$ ... are unique for a given vector, $|\psi\rangle$. The condition for a set of vectors to form a basis is that no basis vector can be written as a combination like (4.1.1) of the remaining basis vectors (i.e., the basis vectors are linearly independent). Also, there must be sufficiently many basis vectors to permit all vectors in the space to be written like (4.1.1) (i.e., the basis is complete).

Every vector space has a basis. The provability of this depends, yet again, on the variant of axiomatic set theory one chooses to adopt. It is provable within (the usually assumed) Zermelo–Fraenkel-set-theory-with-the-axiom-of-choice (ZFC theory), see chapter 1.

The minimum number of basis vectors required for completeness is the dimension of the vector space. This may be a finite number or infinite. As I have written (4.1.1), with the basis vectors labelled by the integers, 1, 2, 3…, if this extends to infinity it would be a denumerable infinity. This corresponds to measurements with a discrete set of possible outcomes, $q_1, q_2, q_3,$ …. (in which case the observable is said to have a discrete "spectrum"). However, observables may also have a continuous spectrum, corresponding to a vector space with an indenumerably infinite dimension (specifically a cardinality of aleph-one, see chapter 1). In this case the notation used in (4.1.1) would be inappropriate and the basis vectors would need to be rewritten with a continuous label, such as x, rather than an integer index. The sum in (4.1.1) then becomes an integral, so an arbitrary vector would then be written $\int f(x)|q(x)\rangle \, dx$. However, I shall avoid continuous spectra in this book, though they are the norm in conventional expositions of quantum mechanics, not least because space is (or is assumed to be) a continuum.

4.1.2 Hilbert Space

A Hilbert space is a vector space with an inner product, i.e., for every ordered pair of vectors, say $(|\psi\rangle, |\phi\rangle)$, there is a complex number which is written in Dirac notation as $\langle\phi|\psi\rangle$. The inner product taken in the opposite order is the complex conjugate, $\langle\psi|\phi\rangle = \langle\phi|\psi\rangle^*$. This means that the inner product of a vector with itself is necessarily real, as $\langle\psi|\psi\rangle = \langle\psi|\psi\rangle^*$. However, $\langle\psi|\psi\rangle$ is also required by the conditions of Hilbert space to be positive definite for all non-zero vectors. The "norm" (roughly the "magnitude") of any vector is defined as $\|\psi\| = \sqrt{\langle\psi|\psi\rangle}$. The inner product is also required to be linear in the sense that,

$$\langle \omega[a|\psi\rangle + b|\phi\rangle] = a\langle \omega|\psi\rangle + b\langle \omega|\phi\rangle$$

Two vectors such that $\langle \psi|\phi\rangle = 0$ are said to be orthogonal. This generalises the concept of orthogonality in physical space, the inner product generalising the dot product (or scalar product) of vectors in ordinary (Euclidean) space.

A set of vectors which are all mutually orthogonal and also all have unit norm is said to be orthonormal. The Hilbert space inner product therefore provides a definition of an orthonormal basis, which generalises the idea of a Cartesian coordinate system.

4.1.3 Operators and Eigenvalues

I shall be lazy and refer to "operators" when, throughout, I shall mean linear operators. An operator maps a Hilbert space vector into another vector in the same Hilbert space. An operator, Q, is linear if it respects the following conditions: $Q(|\psi\rangle + |\phi\rangle) = Q|\psi\rangle + Q|\phi\rangle$ and $Qa|\psi\rangle = aQ|\psi\rangle$. Also two linear operators, P and Q, respect $(P + Q)|\psi\rangle = P|\psi\rangle + Q|\psi\rangle$.

For a given operator there will generally be some special vectors, $|\lambda_i\rangle$, such that the action of the operator on these special vectors is just to multiply them by some number (in general, complex), and I have labelled the vector itself with this number, i.e.,

$$Q|\lambda_i\rangle = \lambda_i|\lambda_i\rangle \qquad (4.1.2))$$

These numbers, λ_i, are called the operator's eigenvalues, whilst the vectors, $|\lambda_i\rangle$, are the corresponding eigenvectors. Eigenvalues are of central importance in quantum mechanics because they are the possible values taken by observable physical quantities.

As $Q|\psi\rangle$ is a vector we can form the inner product with another vector, i.e., $\langle \phi|Q|\psi\rangle$. This is also known as the "matrix element of Q between states ϕ and ψ". The adjoint, or Hermitian conjugate, operator corresponding to Q is written Q^+ and is defined such that, for every pair of vectors,

$$\langle \psi|Q^+|\phi\rangle \equiv \langle \phi|Q|\psi\rangle^* \qquad (4.1.3)$$

An operator which equals its own adjoint is said to be Hermitian.

Hermitian operators play a key role in quantum mechanics because their eigenvalues are real and their eigenvectors form an orthonormal basis (for the proof see chapter 2 of *The Unweirding*).

4.1.4 Interpretational Principles of Quantum Mechanics

The formalism of Hilbert space is all very well, but what does it mean physically? This is specified by the following foundational Principles,

[1] Pure quantum states are represented by Hilbert space rays, i.e., vectors with unit norm.

[2] Physical observables are represented by Hermitian operators.

[3] The possible outcomes of a measurement of an observable $\hat{Q}$ are the eigenvalues of $\hat{Q}$, i.e., $q_1, q_2, q_3, \ldots$ (which are necessarily real because $\hat{Q}$ is Hermitian).

It follows that the eigenvectors of $\hat{Q}$ form an orthonormal basis, and hence we can express an arbitrary pure quantum state in the form of (4.1.1), i.e., $|\psi\rangle = \sum_i a_i |q_i\rangle$. The last two Principles are,

[4] (Born's Probabilistic Rule): For a pure quantum state $|\psi\rangle$ the probability that the outcome of a measurement of observable $\hat{Q}$ will be q_i is $|a_i|^2$.

[5] (Collapse of the State): If the measurement outcome is q_i then the system is left in state $|q_i\rangle$ after the measurement.

The last Principle causes endless controversy and lies at the heart of the measurement problem of quantum mechanics, a problem which still has no satisfactory resolution that has achieved consensus agreement.

Principle [4] renders quantum mechanics irreducibly indeterminate. Classical physicists were most uncomfortable with this aspect of quantum mechanics and many tried to make measurement outcomes deterministic by introducing hidden variables in one form or another. However, hidden variables are now firmly refuted and quantum mechanics really does stand revealed as irreducibly indeterminate.

Readers will note my repeated use of the phrase "pure quantum state". Be aware that these are rather special states of a system. An arbitrary physical system cannot be assumed to be in a pure state, and hence cannot be described by a Hilbert space vector. Failure to realise this leads to nonsense like Schrodinger's cat. In general, the description within quantum mechanics of an arbitrary physical system is as a so-called "mixed state". I shall avoid any discussion of mixed states in this book, beyond just this. "Large" systems can be hard to coax into a pure state. "Large" here really means merely systems with many degrees of freedom, which might still be molecular in size. Most systems which start in a pure quantum state (especially those with many degrees of freedom) tend to degenerate into a mixed state due to interactions with the environment. This is known as "decoherence". (There are exceptions. Some pure quantum states are indefinitely stable, e.g., atomic electron orbits, for example.)

All the above issues are addressed in detail in *The Unweirding*.

4.1.5 Multiple Systems

One final bit of formalism is needed to appreciate how quantum computers work. This is the manner in which the pure state of a system comprising N similar subsystems can be represented. In this case the basis for the description of the combined system is written,

$$|q_i\rangle \otimes |q_j\rangle \otimes |q_k\rangle \otimes \dots$$

The symbol $\otimes$ does not imply any computation analogous to ordinary multiplication and so the above "tensor product", as it is strictly called, is often written simply as the juxtaposition $|q_i\rangle|q_j\rangle|q_k\rangle$ ... However, the tensor product is slightly more than a mere ordered set because it is required to obey

$$a_i|q_i\rangle \otimes a_j|q_j\rangle \otimes a_k|q_k\rangle \dots = a_i a_j a_k |q_i\rangle \otimes |q_j\rangle \otimes |q_k\rangle \dots$$

And this makes perfect sense when it is written, by analogy with an ordinary numerical product, as $a_i|q_i\rangle a_j|q_j\rangle a_k|q_k\rangle \dots = a_i a_j a_k |q_i\rangle|q_j\rangle|q_k\rangle$ .... Other than that technical nicety the general state of N similar subsystems can be written,

$$\sum_{i,j,k,\dots} A_{ijk\dots} |q_i\rangle|q_j\rangle|q_k\rangle \dots \qquad (4.1.4)$$

Note that this state of N subsystems cannot, in general, be expressed as a product of N single-system states, a mathematical fact that gives rise to the phenomenon of entanglement (for more of which see *The Unweirding*).

4.2 Computation: Classical and Quantum

Classical computation is based on physical systems in binary states, or which can be interpreted with extremely high fidelity in a binary fashion. The two states can be distinguished by labelling them as 0 and 1 (and will, indeed, be interpreted as standing for such binary digits). The physical nature of the system need not concern us. The two states might be the presence or absence of an electrical charge, or the presence or absence of a certain voltage, or whether a particle is spin up or spin down. In a conceptual "billiard ball computer" (Fredkin and Toffoli, 1982) the two states are the presence or absence of a ball. A concatenation of several of these "bits" (<u>bi</u>nary di<u>g</u>its) can then be used to represent integers (below a certain maximum size) or, with limited precision, real numbers.

Readers will be familiar with the basic idea of digital computers. Physical processes can be carried out within the device which reproduce the action of logical operations. For this purpose the states labelled 0 and 1 are interpreted as having the truth values "false" (F) and "true" (T) respectively. At the simplest level, the physical processes which implement the logic on just a few bits are called "gates". Gates may operate on just one bit, or on two or three or more bits in parallel. The action of a gate can be defined by a truth table, or equivalently by a matrix.

Consider the gate representing "NOT" which operates on a single bit. If the input bit is 0 (i.e., F) then the gate outputs 1 (i.e., T), and vice-versa. This can be written as the truth table,

Input	Output
0	1
1	0

Alternatively, if we represent 0 (or F) by the column matrix $\begin{pmatrix} 1 \\ 0 \end{pmatrix}$ and 1 (or T) by $\begin{pmatrix} 0 \\ 1 \end{pmatrix}$, then the operation of the NOT gate is represented by the matrix $\begin{pmatrix} 0 & 1 \\ 1 & 0 \end{pmatrix}$ because we then have,

$$\begin{pmatrix} 0 & 1 \\ 1 & 0 \end{pmatrix}\begin{pmatrix} 1 \\ 0 \end{pmatrix} = \begin{pmatrix} 0 \\ 1 \end{pmatrix} \quad \text{and} \quad \begin{pmatrix} 0 & 1 \\ 1 & 0 \end{pmatrix}\begin{pmatrix} 0 \\ 1 \end{pmatrix} = \begin{pmatrix} 1 \\ 0 \end{pmatrix}$$

The reader will see how this immediately reflects the structure of the formulation of quantum mechanics. The state of the physical system is being represented by a "vector", i.e., a column matrix, whilst the physical operation performed on the state is represented by an "operator", i.e., a square matrix, which maps one vector into another. If the bits were physically instantiated as pure binary quantum states, then we can write,

$$|F\rangle = \begin{pmatrix} 1 \\ 0 \end{pmatrix} \quad \text{and} \quad |T\rangle = \begin{pmatrix} 0 \\ 1 \end{pmatrix}$$

The gate is then manifest physically by a process $\hat{U}$ which has the action,

$$\hat{U}|F\rangle = |T\rangle \quad \text{and} \quad \hat{U}|T\rangle = |F\rangle$$

So, what's the difference between a classical computer and a quantum computer? The key difference is that the most general state of the system representing one "bit" in a quantum computer is not either $|F\rangle$ or $|T\rangle$ but any linear combination: $\alpha|F\rangle + \beta|T\rangle$, where α, β are any complex numbers (subject to normalisation, $|\alpha|^2 + |\beta|^2 = 1$).

$\alpha|F\rangle + \beta|T\rangle$ is the "qubit" (the quantum bit).

Note that this is non-trivial. It is not the same as a classical system that can be in a continuum of states, defined by potentially measurable α, β. A measurement of the truth-value of the qubit will always produce either T or F as the outcome. The qubit is still a two-state device, just as was the classical bit. But its state, prior to measurement, is in general a superposition of $|F\rangle$ and $|T\rangle$. However, after measurement it will be left in state $|F\rangle$ or $|T\rangle$ with probabilities $|\alpha|^2$ and $|\beta|^2$ respectively.

And what use might this be, you could reasonably ask? Making the truth value of a state uncertain would seem to be a retrograde step.

The potential utility of a quantum computer, namely hugely increasing computational speed, relates to the opportunity for massively parallel processing it offers. This can be seen by reverting to the interpretation of the basis states as binary digits. A string of N such qubits can represent 2^N integers. But the combined quantum state of the N qubits, provided it is a pure state, is of the form of a tensor product, as given by (4.1.4). That is, by $\sum_{i,j,k,...} A_{ijk...} |q_i\rangle|q_j\rangle|q_k\rangle$ .... where each state in the product is $|0\rangle$ or $|1\rangle$, but the sum will, in general, include both these states for every qubit, and potentially in every combination.

So, whilst we can interpret a state $|1\rangle|0\rangle|0\rangle|1\rangle$ (say) as representing the integer 9, what number does $\sum_{i,j,k,...} A_{ijk...} |q_i\rangle|q_j\rangle|q_k\rangle$ ... represent? If the number of subsystems (i.e., qubits) is N, then every integer from 0 to $2^N - 1$ occurs in the sum (potentially with different weightings, but perhaps the same weighting if all the $A_{ijk...} = 2^{-N/2}$).

Consider a program to implement some computation, which, in a quantum computer is implemented by an operator $\hat{U}$. Suppose we need to run the program for every possible input value from 0 to $2^N - 1$. A classical, single-processor computer will therefore take a time $2^N t$ to complete the task, where t is the time for a single run for one input. In contrast, the potential benefit of the quantum computer is that it could run just once, taking as input the N-qubit state $\sum_{i,j,k,...} 2^{-N/2} |q_i\rangle |q_j\rangle |q_k\rangle$ .... By that means the computation is performed simultaneously on all inputs from 0 to $2^N - 1$ in a single run.

The potential for computational speed-up is therefore of order of 2^N, which is enormous for large N. This is why the challenge to develop quantum computers which outperform classical computers hinges crucially on the number of qubits which the quantum computer can process in this way.

There are, of course, enormous technical challenges to producing a quantum computer which delivers this promised performance – even under ideal laboratory conditions. The biggest problem is decoherence.

Recall that quantum superposition is the defining feature which makes quantum computers "quantum". But superposition applies only to pure quantum states. (I'm simplifying a little here). Unfortunately, pure quantum states are generically prone to degenerating into a state in which the superposition is lost. The cause is interaction with the environment. For example, if the qubit state $\alpha|0\rangle + \beta|1\rangle$ decohered it would become either $|0\rangle$ or $|1\rangle$. A quantum computer operating on a fully decohered input state would, therefore, be running the computation for just a single input – the same as a classical computer, and therefore ceases to offer any advantage. Moreover, the larger the number of qubits, the harder it becomes to maintain coherence (for the reason why, see *The Unweirding*).

Even if you did manage to maintain the coherence of your quantum state for long enough to carry out the computation, there is then the problem of reading the result. What the quantum computer delivers is a quantum state. What you want is an answer – such as "yes" or "no" or "42" or the input string required to reproduce a given cryptographic hash value. One has no choice but to make a measurement of the final state (although it could be subject to processing of some sort first). But quantum measurement is an irreversible process: the state is collapsed by the measurement and you can't get the same state back again. So you'd better be sure the measurement will provide you with the desired answer. Details of the strategies that have been explored as regards "readout" would take us beyond the purposes of this book; see a specialist text or web site on quantum computing.

4.3 The Square-Root of NOT

It has taken many pages to set the scene but we are now ready for the square-root of NOT to put in an appearance.

In standard logic there is no logical operator which, if applied twice, is equivalent to the operation NOT. The reason is that standard logic is bivariate, i.e., the truth value of every statement is T or F. Suppose the proposed operator is denoted Q, so that $Q^2 \equiv NOT$. We therefore require that $Q^2T = F$ and $Q^2F = T$. But what values can QT and QF take consistent with this if they must be either F or T? Clearly this is impossible, for if $QT = T$ then $Q^2T = Q(QT) = QT = T$, which is not what we require. On the other hand if $QT = F$ then $Q^2T = Q(QT) = QF$ so we require $QF = F$ but in that case $Q^2F = Q(QF) = QF = F$ which is not what we require.

For there to be such a thing as the square-root of NOT, there must be some value that QT and QF can take which is neither T nor F. One might be driven to think of a trivalued logic, say, in which there is "middle ground" between T and F. Even that would not be enough, though. We would need two distinct "middle" truth values, call them P and R. Then we could define,

$$QF = P \qquad QP = T \qquad QT = R \qquad QR = F$$

which then give us $Q^2T = F$ and $Q^2F = T$, as required. So this would be a rendering of $Q = \sqrt{NOT}$ using four distinct truth values.

The obvious problem with this scheme is the interpretation of these purely formal (and artificially invented) "truth values", P and R. I will consider this and related issues further in §4.4.

But now I wish to point out that quantum computation comes ready-made with a solution that, on the one hand, provides intermediate states analogous to P and R, but, rather miraculously, without having to recognise as genuine "truth values" anything other than F and T. The secret is, of course, the key attribute of quantum mechanics: that states can come as superpositions. The generic qubit is $\alpha|F\rangle + \beta|T\rangle$, which is neither T nor F but will become one or the other upon measurement. Thus, we have an intermediate state which avoids being either T or F, but which still conforms to being either T or F when measured. And so…in quantum computation $\sqrt{NOT}$ exists.

Expressing the basis of the qubit as $|F\rangle = \begin{pmatrix} 1 \\ 0 \end{pmatrix}$ and $|T\rangle = \begin{pmatrix} 0 \\ 1 \end{pmatrix}$ the operator representing $\sqrt{NOT}$ is,

$$\sqrt{NOT} = \frac{e^{i\frac{\pi}{4}}}{\sqrt{2}} \begin{pmatrix} 1 & -i \\ -i & 1 \end{pmatrix} \tag{4.3.1}$$

So we get,

$$\sqrt{NOT}\begin{pmatrix} 1 \\ 0 \end{pmatrix} = \frac{e^{i\frac{\pi}{4}}}{\sqrt{2}} \begin{pmatrix} 1 \\ -i \end{pmatrix} \quad \text{and} \quad \sqrt{NOT}\begin{pmatrix} 0 \\ 1 \end{pmatrix} = \frac{e^{i\frac{\pi}{4}}}{\sqrt{2}} \begin{pmatrix} -i \\ 1 \end{pmatrix} \tag{4.3.2}$$

And then a repeat application of $\sqrt{NOT}$ gives,

$$NOT\begin{pmatrix} 1 \\ 0 \end{pmatrix} = \frac{e^{i\frac{\pi}{2}}}{2} \begin{pmatrix} 0 \\ -2i \end{pmatrix} = \begin{pmatrix} 0 \\ 1 \end{pmatrix} \quad \text{and} \quad NOT\begin{pmatrix} 0 \\ 1 \end{pmatrix} = \frac{e^{i\frac{\pi}{2}}}{2} \begin{pmatrix} -2i \\ 0 \end{pmatrix} = \begin{pmatrix} 1 \\ 0 \end{pmatrix}$$

recalling that $e^{i\frac{\pi}{2}} = i$. More simply, by squaring the matrix in (4.3.1) we get immediately,

$$\left(\sqrt{NOT}\right)^2 = \left(\frac{e^{i\frac{\pi}{4}}}{\sqrt{2}} \begin{pmatrix} 1 & -i \\ -i & 1 \end{pmatrix}\right)^2 = \frac{e^{i\frac{\pi}{2}}}{2} \begin{pmatrix} 0 & -2i \\ -2i & 0 \end{pmatrix} = \begin{pmatrix} 0 & 1 \\ 1 & 0 \end{pmatrix} = NOT$$

The important point to appreciate is that (4.3.1) represents a gate which is physically realisable. So there is a piece of hardware which, if a qubit in state $|F\rangle$ is fed into it will produce the first of the states (4.3.2). And if this is fed into the same hardware for a second time will produce $|T\rangle$.

There need not be anything exotic about the hardware required to do this. Indeed, it might be quite prosaic. For example, if the basis states were realised as the up and down spin states of a spin ½ particle, then it is clear that the NOT operation – which transforms one into the other – can be realised simply by a 180° rotation. It follows that $\sqrt{NOT}$ would be realised simply as a 90° rotation. And that is, in fact, exactly what the matrix (4.3.1) is: the rotation matrix for a 90° rotation acting on a spin ½ state.

4.4 The Law of the Excluded Middle

A fierce debate raged amongst logicians and foundational mathematicians in the nineteenth and early twentieth centuries. This centred around the issue we touched upon in the last section: whether statements can have a truth value which is neither "true" nor "false". The matter can be expressed as follows: for an arbitrary statement, A, is "A OR (NOT A)" always true? If we assume that all sentences are either true or false, then this follows immediately. In other words, if we assume the so-called Law of the Excluded Middle, namely that T and F are the only truth values, there is nothing in-between, then "$A\ OR\ (NOT\ A)$" always has truth value T. (Here OR is the inclusive OR, so that $A\ OR\ B$ is true if either or both of A and B are true).

The intuitionist and constructivist schools of mathematical thought rejected this. They reject the Law of the Excluded Middle (though they do not necessarily posit what might be "in the middle"). Adopting their alternative perspective has major repercussions on issues of mathematical existence and the availability of proof methodologies. For example, in conventional mathematics, it is common to present proofs that an entity (e.g., the solution to an equation) exists, without having explicitly constructed that entity (e.g., without having found the solution to the equation). But the hard-line constructivist position is that such proofs are invalid unless the entity has been explicitly constructed.

Perhaps the most serious consequence of adopting the intuitionist/constructivist position is that *Reductio Ad Absurdum*, that familiar workhorse of mathematical proof, is no longer acceptable. This is because *Reductio Ad Absurdum* requires "$A\ OR\ (NOT\ A)$" to be identically true. The logic of *Reductio Ad Absurdum* is to show that assuming $NOT\ A$ leads to a contradiction, and hence that $NOT\ A$ is not true. The Law of the Excluded Middle then implies that $NOT\ A$ is false, and the identity "$A\ OR\ (NOT\ A)$" tells us that A must be true. So, *Reductio Ad Absurdum* requires the Law of the Excluded Middle (which implies $A\ OR\ (NOT\ A)$).

This is perhaps why, despite the debate between the intuitionist/constructivist schools of thought and the more conventional approaches having never been strictly resolved, nevertheless, in practice, mathematics goes on as before: mathematicians would be reluctant to forego *Reductio Ad Absurdum*. In Russell and Whitehead's *magnum opus*, the Principia Mathematica, the identical truth of "$A\ OR\ (NOT\ A))$" appears as a theorem, but only because it is derived from axioms with which the intuitionist would equally disagree.

In natural language it is easy to invent sentences which are neither true nor untrue. The most trivial example is a nonsense sentence, such as "mome raths outgrabe". True or false? More serious are the self-referential sentences. Sentences like the Liar Paradox, "I am lying" or "This sentence is false", cannot consistently be assigned a truth value T or F. In both cases, if the sentence is true then it is also false, and vice-versa.

[The sentence "I always lie", on the other hand, cannot consistently be assigned as true but can consistently be assumed as false. Consequently, assuming the Law of the Excluded Middle, making the statement "I always lie" logically implies that I sometimes tell the truth and also sometimes lie].

An equally paradoxical sentence is "This sentence is true". It can consistently be regarded as true. But it can equally consistently be regarded as false. Unlike the Liar Paradox, which cannot be either true or false, "This sentence is true" can be either true or false. There is an analogy here with a qubit before and after measurement. Before measurement the generic state of the qubit is $\alpha|F\rangle + \beta|T\rangle$ and is therefore neither F nor T. But after measurement it can become F or T, and we will not know which until the indeterminate measurement is made.

The unworkability of logic based on unrestricted natural language, in which self-reference is permitted, reaches its apotheosis in Curry's Paradox. In symbolic notation this invites us to consider $A \equiv (A \Rightarrow B)$. In words: the sentence A is defined as being "A implies B". All we need do to show that A is true is to show that **if** it were true then B would necessarily be true, because that is what "A implies B" means. But if A were true, then we conclude that "A implies B" is true, because that is what A asserts. But since we now have that "A implies B" is true then, under the precondition that A itself is true, we conclude that B is true. And this establishes, in the manner stated above, that A is indeed true. And hence B is true.

The snag is that B can be any statement whatsoever.

If allowed in a system of thought, this $A \equiv (A \Rightarrow B)$ monster would allow us to prove anything, including its negation. Such a possibility must, therefore, be outlawed if logic is not to collapse in a heap of rubble. Alternatively, if you wish to prove that Boris Johnson is a chipmunk, this is the logic for you.

More seriously, this illustrates why the foundations of logical deduction, including the foundations of mathematics, can only attain full cogency by restricting their sphere of discourse. In set theory this was famously illustrated by Russell's paradoxical "set of all sets which do not contain themselves", which contains itself only if it doesn't, and vice-versa. Russell's Theory of Types was invented to protect mathematics from such horrors by excluding them, the resulting basis of mathematics being presented in Russell and Whitehead's *Principia Mathematica*. The more developed system which excludes such logical nasties as Russell's Paradox, Curry's Paradox, etc., is our old friend Zermelo–Fraenkel set theory (see §1.9).

A question arises from this pragmatic approach, however: by limiting our sphere of discourse, what do we excise from our worldview? I risk making a few remarks in the next section which go beyond the strict subject matter of this book.

4.5 Deeper Thoughts

Here I dip my toe in the deepest of waters. I shall not venture far. Readers must bear in mind that, in this section especially, I stray well beyond the areas of competence that I could reasonably claim.

The discussion of §4.4 has shown that the power of mathematics is bought at the cost of limiting its sphere of discourse, namely by excluding those features that would cause trouble. Physical science has a similar characteristic. The power of physical science rests upon the imposition of objectivity. It is easy to overlook this because it has become second nature in the scientific community. But the rigorous exclusion of the subjective from physical science is part of the empirical method, which, together with "conjecture and refutation", provide such a powerful deductive system. This is another instance of how progress in an area of knowledge can rest upon the limitations imposed on the allowed sphere of discourse.

A more obvious example of such deliberate limiting is that physical science starts by restricting its attention to matter. The success of science has given rise to a widespread belief in materialism, namely that matter is all there is. The circularity of the reasoning is stark. One cannot start by making an assumption and then, after tortuous reasoning, end by concluding the assumption has been proved (unless, perhaps, one is a fan of Curry's Paradox). Thus, materialism is not demonstrated by physical science because physical science founds its approach upon excluding the non-material. Materialism is therefore a faith, not a scientific conclusion – an observation which will discombobulate most of its adherents who tend to believe that materialism is the very absence of faith.

As a physicist I must confess that materialism is a plague which we released upon the world. But physicists themselves are now recovering from the disease, whilst leaving much of the rest of the world still ailing from it. It was always rather mad to imagine that the ultimate source of absolutely everything was tiny little billiards balls of lumpen 'stuff'. Had this turned out to be so, it would have been like something dreamed up by Terry Pratchett. But it has not – inevitably. Instead, the closer physicists have looked at matter, the more it disappears before one's eyes.

Let me emphasise the point. The essence of matter is its mass (including the mass-equivalent of energy, so light has mass in this sense, despite its zero "rest mass"). Mass therefore characterises everything that is material in nature, even electromagnetic radiation, neutrinos, gluons, and so on. Be aware of this then: that more than 99% of the mass of ordinary stuff – your desk, your mobile, your head – actually consists of the field energy of gluons. And the bit of mass which is left over arises from the interaction between particles which would otherwise have zero rest mass and the all-pervasive Higgs field.

So, mass comes down to the gluon and Higgs fields, plus the mass equivalent of other quantum field energies. It is hard to convey fully how unlike lumpen stuff are these entities. Any understanding physics has of them is entirely via quantum field theory. What then are quantum fields? They are rather horribly ill-conditioned infinite sums over Hilbert space operator valued entities, the creation and annihilation operators. A quantum field is not a "thing". It is a mathematical capturing of the fizzing potentiality inherent in the vacuum, particles of all sorts constantly appearing from nothing and disappearing again as quickly.

A quantum field is less a noun than a verb: it acts upon other things. But, as all possible other "things" are ultimately quantum fields, these other "things" are also not really things. Confused? Me too.

Are quantum fields the ultimate truth? I doubt it, but on the other hand the development of physics suggests they are closer to the truth than what came before – namely the almost Discworld idea of tiny billiard balls.

The message is that science makes a Faustian bargain. In return for knowledge and power over the world, the scope of our understanding is restricted. In falsely concluding that "science proves materialism" one not only falls into an obvious logical error but also reaps the reward of all Faustian bargains: nihilism. It is as well to bring back to mind that this has resulted from logical error.

The vanishing away of matter the closer one looks at it puts me in mind of the Buddhist principle of *sunyata*, or emptiness. There is a famous dispute within Buddhist doctrinal methodology between the Svatantrika and the Prasangika schools of thought. I shall not attempt an explanation of the details of the dispute, of which I can have no great understanding as an unenlightened person. However, the Prasangika school is happy for a Buddhist logician to use a form of *Reductio Ad Absurdum* (which they call *prasanga*) to show how an essentialist position leads to false conclusions. In contrast, the Svatantrika school believes that the Buddhist must also put forward a concrete thesis of his own, not merely rely upon negation. The Prasangika rejection of the Svatantrika position was based on the belief that any Buddhist making positive assertions about the conventional world was committed to the existence of an illusion. My understanding is that the Prasangika is the more profound understanding and has gained general acceptance.

The Prasangika position crucially involves non-affirmative negation. Not only is the world empty, but that must not be interpreted as putting some substance called "emptiness" in its place: the emptiness is itself empty. (I write the words but have no idea what I'm talking about). However…

Under non-affirming negation, if one were to assert the falsity of the statement "the world of mundane physical objects truly exists" that would not imply either that the world of mundane physical objects does not exist (solipsism) nor that something else exists in the place of the world of mundane physical objects.

Similarly, the falsity of the statement "humans have an immortal soul" would not imply that "humans are totally annihilated at death". This example is key in Buddhism as it poses the dichotomy between eternalism and nihilism, and hence between theistic religions and irreligious atheism. Buddhism rejects the dichotomy and embraces the Middle Way, in which neither of those statements is true. Hence, at least in a rough-and-ready sense, Buddhism rejects the Law of the Excluded Middle and holds that *A OR (NOT A)* is not always true when the sphere of discourse is allowed to embrace the ultimate metaphysical mysteries.

The reader may be puzzled that the Prasangika school advises the use of *prasanga*, essentially *Reductio Ad Absurdum*, when it also rejects the Law of the Excluded Middle upon which *Reductio Ad Absurdum* rests. The resolution of this puzzle is that the Prasangika recommend the use of *prasanga* only within the context of refuting the essentialist position, i.e., in opposing a body of opinion in which *Reductio Ad Absurdum* can be used. The enemy's weapon is used against it, so to speak. The ultimate Prasangika position, however, is to refute the Excluded Middle, and hence to refute the method of *Reductio Ad Absurdum*, paradoxically by deployment of *Reductio Ad Absurdum* to do so.

4.6 References

Fredkin, Edward; Toffoli, Tommaso (1982), "Conservative logic", *International Journal of Theoretical Physics*, **21** (3–4): 219–253

5

The Square Root of Geometry

Here we meet yet another example of insisting that an impossible square root be made possible by suitably extending our concept of "number". Yet again this process yields remarkable fecundity. In the case of Paul Dirac, this approach led to the most stunning prediction in physics of all time: the purely theoretical prediction of antimatter. This example is particularly deep because it leads to the existence of matter being intimately related to a sort of "square root of geometry", in a sense to be explained herein.

Here we arrive at the most miraculous impossible square root of them all, and one with a most unexpected physical import. Moreover, the associated prediction was experimentally confirmed with remarkable rapidity. The prediction I refer to, of course, was Dirac's prediction of the existence of antimatter. There was no intimation of such a thing – neither theoretical nor experimental – before Dirac pulled this stupendous rabbit out of a hat. The hat in question was the relativistic expression for energy in terms of mass and momentum. In this chapter I describe how the trick was performed – and, you guessed it, the crucial step was in refusing to accept that a particular square root could not be done. Instead, as we have seen many times before, one merely extends one's horizons to embrace a realisation of the previously impossible.

Before launching into the derivation, which is not difficult, it is worth pausing to note that, whilst the prediction of antimatter is the headline-grabbing achievement, there is actually something more profound still lurking in Dirac's discovery. It is that, in a sense to be elucidated in this chapter, matter itself is the square root of geometry.

5.1 Dirac's Miraculous Square-Root

The repeated theme of this book is that by refusing to accept that a square root is impossible, extensions of one's existing conceptions of 'number' are created. When applied to physics, the result can be a startling prediction about the external world. The supreme example is Dirac's prediction of antimatter (1928 – 1931). In 1926 Schrodinger had published his famous equation, which, in essence, is just the statement that total energy is the sum of kinetic energy and potential energy – in non-relativistic form,

$$E_{nr} = \frac{p^2}{2m} + V \tag{5.1.1}$$

where p is a particle's momentum, m its mass, and V its potential energy. The subscript on E_{nr} denotes that this is the non-relativistic energy (which, in particular, excludes the rest-mass energy). The Schrodinger equation proper is obtained by replacing E_{nr} and p in (5.1.1) with certain differential operators which represent the operator-observables within quantum mechanics. Specifically,

$$E_{nr} \rightarrow i\hbar \frac{\partial}{\partial t} \quad \text{and} \quad \bar{p} \rightarrow -i\hbar \bar{\nabla} \quad \text{or} \quad p^2 \rightarrow -\hbar^2 \nabla^2 \qquad (5.1.2)$$

where t is time and $\bar{\nabla}$ is the operator sometimes denoted "grad". These operators have to act on something, and in the Schrodinger picture this thing is the wavefunction, ψ. Hence the Schrodinger equation is just

$$i\hbar \frac{\partial}{\partial t} \psi = \left(-\frac{\hbar^2}{2m} \nabla^2 + V \right) \psi.$$

(Actually, Schrodinger's 1926 equation was not in this time-dependent form, but let that historical detail pass. The equation was probably first written in this form by Dirac, but not published until Schrodinger did so).

Don't worry about (5.1.2) or the Schrodinger equation if you are not familiar with differential operators or differential equations, no knowledge of them will be required. The only thing to appreciate is that the Schrodinger equation, i.e., (5.1.1) after substitution of (5.1.2), is of first order in time, that is it is linear in E_{nr} and hence linear in the time derivative, $\frac{\partial}{\partial t}$.

This is of profound significance when the Schrodinger equation is applied to atomic spectra, for which the potential energy is the electrostatic Coulomb energy between the negative electrons and the positive nucleus. It is this mathematical property, being first order in E_{nr} and hence in time, which leads to there being a lowest energy state of finite energy, i.e., a ground state of atomic orbitals. This is the crucial feature of Schrodinger's 1926 paper (together with the predicted simple dependence of the allowed energy states on a single principal quantum number).

Once it had been realised that atoms consisted of negative electrons orbiting a positive nucleus, there was immediately a conundrum regarding why such electrons – being, apparently, accelerating charges – did not simply radiate away their kinetic energy as was predicted by the classical Maxwell equations of electrodynamics. Why did the electrons not end up collapsing into the nucleus, with virtually divergent (negative) potential energy? But if the Schrodinger equation was postulated as defining the only possible electrons states, this problem was solved because the Schrodinger equation predicted the existence of a lowest energy state of finite energy, the ground state. The electron could not spiral into the nucleus because there was no such state available.

Enter Dirac, whose focus of concern was to derive an equation like Schrodinger's but consistent with (special) relativity. We have already seen in §3.7 how the relationship between energy and momentum changes in relativity. Instead of (5.1.1) we now have, for the case of a free particle with zero potential energy,

$$E^2 = p^2 + m^2 \tag{5.1.3}$$

and the same substitutions in terms of differential operators, (5.1.2), continue to apply. The subscript nr has been dropped from the energy as E is now the full relativistic energy, including the rest-mass energy (as can be seen by replacing p by zero). Note that we are using units in which the speed of light, $c = 1$.

The reader will see the problem: (5.1.3) is now of second order in the differential operator $\frac{\partial}{\partial t}$. This wrecks the principal success of the Schrodinger equation because even free electrons could now be assigned negative energies of arbitrarily large magnitude. The simplest way of seeing this is that (5.1.3), interpreted in terms of ordinary numbers, has two square roots,

$$E = +\sqrt{p^2 + m^2} \quad \text{and} \quad E = -\sqrt{p^2 + m^2} \tag{5.1.4}$$

So, as a free electron can have an arbitrarily large positive energy and momentum, it can therefore also have an arbitrarily large negative energy. A free electron would then be unstable and cascade downwards in energy without limit. This is catastrophic.

The problem would be solved, thought Dirac, if he could find an operator which was the square root of the operator on the RHS of (5.1.3). In other words, Dirac sought an operator $\hat{H}$ such that $\hat{H}^2 = p^2 + m^2$. The problem with this is that, if we confine attention to the complex number field, the Fundamental Theorem of Algebra tells us that $p^2 + m^2$ has a unique factorisation and that this is $p^2 + m^2 = (p + im)(p - im)$. Since these two factors are different, then there is no $\hat{H}$ such that $\hat{H}^2 = p^2 + m^2$.

Undeterred, Dirac took the time-honoured course – and refused to accept that he had to confine himself to the complex number field. He wanted to force a factorisation as a square, thus,

$$p^2 + m^2 = (\alpha_i p_i + \beta m)^2 \tag{5.1.5}$$

where p_i are the three Cartesian components of momentum, and the repeated subscript implies summation over i. The "generalised numbers" α_i, β were going to have to have some odd properties not shared by real or complex numbers in order to make (5.1.5) possible. The idea of non-commutativity was already familiar in quantum mechanics by 1926, from Born and Heisenberg's work (see chapter 3), and from Pauli's 1926 paper on the hydrogen spectrum. So Dirac was led naturally to extend the algebraic properties of the "generalised numbers" α_i, β to be non-commutative amongst themselves, but to commute with p and m. With this in mind, expanding the brackets in (5.1.5) and noting that the repeated induces are dummy summation variables, gives,

$$p^2 + m^2 = (\alpha_i p_i + \beta m)(\alpha_j p_j + \beta m) = \alpha_i \alpha_j p_i p_j + \alpha_i \beta m p_i +$$
$$\beta \alpha_j m p_j + \beta^2 m^2 = \frac{1}{2}(\alpha_i \alpha_j + \alpha_j \alpha_i) p_i p_j + (\alpha_i \beta + \beta \alpha_i) m p_i + \beta^2 m^2$$

Hence, Dirac's initiative works and the RHS becomes identical to the LHS iff the following relationships hold between the "numbers" α_i, β,

$$\alpha_i \alpha_j + \alpha_j \alpha_i = 2\delta_{ij} \qquad \alpha_i \beta + \beta \alpha_i = 0 \qquad \beta^2 = 1 \qquad (5.1.6)$$

where Kronecker's delta, δ_{ij}, is 1 when $i = j$ but otherwise zero. Matrices with complex elements are one way of achieving the non-commutative properties (5.1.6). We therefore seek matrices obeying (5.1.6) in terms of which the energy operator (conventionally called the "Hamiltonian") is given by,

$$\hat{H} = \alpha_i p_i + \beta m \qquad (5.1.7)$$

But recall in §4.1.3 we saw that operators representing physically observable quantities must be Hermitian. Since energy is an observable, the matrices α_i, β must be Hermitian, i.e., they must equal their complex-conjugate transpose. Now the eigenvalues of an operator-squared are the squares of the eigenvalues of the operator (because $Q|\lambda_i\rangle = \lambda_i|\lambda_i\rangle$ implies $Q^2|\lambda_i\rangle = Q(\lambda_i|\lambda_i\rangle) = \lambda_i Q|\lambda_i\rangle = \lambda_i^2|\lambda_i\rangle$). But (5.1.6) tell us that $\beta^2 = \alpha_i^2 = 1$, where 1 is here interpreted as the unit matrix, and the unit matrix only has eigenvalues of 1. Hence, as the matrices α_i, β are required to be Hermitian and hence must have real eigenvalues, and the squares of their eigenvalues are 1, they can only have eigenvalues ± 1.

But it also follows from (5.1.6) that the trace of each of the α_i, β is zero (the trace being the sum of diagonal elements). For example $\alpha_i \beta + \beta \alpha_i = 0$ together with $\beta^2 = 1$ gives $\alpha_i = -\beta \alpha_i \beta$ and the order is irrelevant when taking the trace, i.e., $Tr(ABC) = Tr(BCA)$, so that,

$$Tr(\alpha_i) = -Tr(\beta\alpha_i\beta) = -Tr(\alpha_i\beta\beta) = -Tr(\alpha_i) = 0$$

Since the first equation of (5.1.6) tells us that $\alpha_i^2 = 1$, by symmetry we similarly get $Tr(\beta) = 0$.

The trace of an Hermitian operator is the sum of its eigenvalues (because if its eigenvectors are used as a basis then its diagonal elements are just $\langle\lambda_i|Q|\lambda_i\rangle = \lambda_i$). But since the eigenvalues of α_i, β are ± 1 it follows that the dimension of these matrices must be even so that there is an equal number of +1 and -1 eigenvalues, in order that they might cancel in the sum.

But there are no two-dimensional (i.e., 2 x 2) matrices obeying (5.1.6). The proof of this is of great importance in its own right, and we look at that next. We shall conclude that the smallest dimension of matrix for which (5.1.6) can hold is 4 x 4. This is the origin of the 4-component Dirac spinors, whose essentially 4-component nature points to the existence of antimatter. But first...

5.2 The Most General Two-State Quantum Observable

We have met two-state quantum systems before in §4.2. When interpreted within the context of quantum computation, a two-state quantum system is a qubit. In Dirac notation the most general state of such as system is simply $\alpha|F\rangle + \beta|T\rangle$, where F and T are merely arbitrary labels of two states which are assumed orthogonal and normalised, i.e., they form an orthonormal basis. In matrix notation an arbitrary state can be written simply as $\begin{pmatrix}\alpha\\\beta\end{pmatrix}$ with respect to the chosen basis.

So much for the most general state in a two-state system, but what about the most general observable? These are represented by the most general Hermitian operator – which, in the chosen basis, becomes the most general 2 x 2 Hermitian matrix. In terms of arbitrary real numbers w, n_x, n_y, n_z the most general 2 x 2 Hermitian matrix can be written down immediately,

$$\begin{pmatrix} w + n_z & n_x - in_y \\ n_x + in_y & w - n_z \end{pmatrix} \tag{5.2.1}$$

This can be written,

$$(Q) = w\begin{pmatrix}1 & 0\\0 & 1\end{pmatrix} + n_x\begin{pmatrix}0 & 1\\1 & 0\end{pmatrix} + n_y\begin{pmatrix}0 & -i\\i & 0\end{pmatrix} + n_z\begin{pmatrix}1 & 0\\0 & -1\end{pmatrix} \tag{5.2.2}$$

The unit matrix is not an interesting observable as every vector is an eigenvector of the unit matrix with eigenvalue 1. The three independent Hermitian operators which represent non-trivial observables, up to an arbitrary real multiplicative constant, are,

$$\sigma_x = \begin{pmatrix} 0 & 1 \\ 1 & 0 \end{pmatrix}, \quad \sigma_y = \begin{pmatrix} 0 & -i \\ i & 0 \end{pmatrix}, \quad \sigma_z = \begin{pmatrix} 1 & 0 \\ 0 & -1 \end{pmatrix} \tag{5.2.3}$$

These are known as the Pauli matrices and occur extensively throughout quantum mechanics. In terms of the Pauli matrices the most general observable of a two-state system can be written,

$$(Q) = n_x \begin{pmatrix} 0 & 1 \\ 1 & 0 \end{pmatrix} + n_y \begin{pmatrix} 0 & -i \\ i & 0 \end{pmatrix} + n_z \begin{pmatrix} 1 & 0 \\ 0 & -1 \end{pmatrix} = n_i \sigma_i \tag{5.2.4}$$

Note that these three Pauli matrices do indeed obey the first of Dirac's requirements, i.e., the first equation of (5.1.6),

$$\sigma_i \sigma_j + \sigma_j \sigma_i = 2\delta_{ij} \tag{5.2.5}$$

However there is no forth non-zero 2 x 2 Hermitian matrix, β, which anti-commutes with all three of the σ_i, i.e., the second equation of (5.1.6) cannot be obeyed by any such matrix. This is because the most general Hermitian matrix is $n_i \sigma_i$ and so we would require $n_i \sigma_i \sigma_x + n_i \sigma_x \sigma_i = 0$. But (5.2.5) means that $n_i \sigma_i \sigma_x + n_i \sigma_x \sigma_i = 2n_x$ and hence we would require $n_x = 0$ and, similarly, $n_y = n_z = 0$, and so the only such β would be the zero matrix and hence could not obey the last of (5.1.6), i.e., $\beta^2 = 1$.

Hence there are no 2 x 2 Hermitian matrices which obey the Dirac conditions, (5.1.6). QED.

In passing we note that the Pauli matrices, (5.2.3), also obey the following, so-called "commutation relations",

$$[\sigma_x, \sigma_y] = 2i\sigma_z \qquad [\sigma_y, \sigma_z] = 2i\sigma_x \qquad [\sigma_z, \sigma_x] = 2i\sigma_y \tag{5.2.6}$$

where the "commutator" $[A, B]$ denotes $AB - BA$ and so is zero if A and B commute. The relations (5.2.6) will turn out to be of crucial importance.

5.3 The Dirac Matrices

We now write down in terms of the Pauli matrices a set of 4 x 4 matrices which obey the Dirac conditions, (5.1.6),

$$\alpha_i = \begin{pmatrix} 0 & \sigma_i \\ \sigma_i & 0 \end{pmatrix} \quad \text{and} \quad \beta = \begin{pmatrix} \mathbb{I} & 0 \\ 0 & -\mathbb{I} \end{pmatrix} \tag{5.3.1}$$

where $\mathbb{I}$ is the 2 x 2 unit matrix, $\mathbb{I} = \begin{pmatrix} 1 & 0 \\ 0 & 1 \end{pmatrix}$. The reader should check that all the relationships of (5.1.6) hold.

The matrices (5.3.1) are not the only possible 4 x 4 representation of the conditions (5.1.6). For some purposes alternative representations, corresponding to alternative basis states, may be convenient. However all representations are equivalent to (5.3.1) in the sense that there exists a transformation which transforms one basis into the other, i.e., a matrix U exists such that $\tilde{\alpha}_i = U^{-1}\alpha_i U$ and $\tilde{\beta} = U^{-1}\beta U$. I'll not prove this here but experts will note that this is the same as saying that the homogeneous Lorentz group has only one inequivalent 4-dimensional representation. The implication of this is that we can safely assume the Dirac representation, (5.3.1), and all results of significance will hold in any representation.

5.4 Rotations

The reason why the Pauli matrices are ubiquitous in quantum mechanics is not only because they form the most general two-state observable, (5.2.4), but also because of their fundamental connection to rotations (and hence to angular momentum, for which see *The Unweirding*). An historical perspective is enlightening here, and it will explain why I initially referred to Dirac's seeking of suitable α_i, β to make (5.1.5) work as a quest for "generalised numbers", though we have ended up representing α_i, β as matrices. Of course, it was by analogy with how the number system has historically been extended by insisting that impossible square roots be made possible. But, as it happens, it is also consistent with the history behind the mathematical description of rotations in 3D space.

If we make a very slight modification by defining the new quantities I, J, K by $I = -i\sigma_x, J = -i\sigma_y, K = -i\sigma_z$ the reader can readily confirm that these quantities have the algebraic properties,

$$I^2 = J^2 = K^2 = IJK = -1 \tag{5.4.1}$$

These are the defining properties of the quaternions, as discovered by William Rowan Hamilton and carved by him into the stone of Brougham Bridge in Dublin on 16th October 1843, immediately upon realising their significance.

Hamilton had sought the quaternions for some time. His motivation was to find an algebraic means of implementing rotations in three-dimensional space. In the two-dimensional plane, a position can be represented by a complex number, and multiplying by the complex number $e^{i\theta}$ causes an anticlockwise rotation of that position about the origin by θ. Hamilton had been looking for a corresponding means of representing positions and rotations in three-dimensional space. The quaternions are the answer. Thus, a position in 3D space can be represented by quaternion $p = xI + yJ + zK$ so that the basis quaternions, I, J, K act as unit vectors. A 3D rotation by angle θ about an axis defined by unit vector $\hat{n}$ is then implemented by the quaternion $q = exp\{\theta(\hat{n}_x I + \hat{n}_y J + \hat{n}_z K)/2\}$, the quaternion representing the rotated position being $\tilde{p} = qpq^{-1}$.

[Aside: the reader will note that, if the magnitude of the rotation, θ, is absorbed into the unit vector $\hat{n}$, i.e., if we put $\bar{n} = \theta \hat{n}$, then the quaternion, q, which implements the 3D rotation is just $q = exp\{-iQ/2\}$ where Q is any two-state observable, (5.2.4)].

The non-commutative property of quaternions is not incidental, it is essential for quantities to represent 3D rotations, as Hamilton required, because 3D rotations are themselves non-commutative. If the reader has never checked this for himself, I suggest you do so now. Take an object and perform two rotations about different axes in 3D space. Now carry out those rotations in the reverse order. In general, the final orientation of the object will be different.

Hamilton and his immediate followers thought of quaternions as a further extension of the number system. Just as the reals extend the rational numbers, and the complex numbers extend the reals, so the quaternions extend further to a type of 'number' with four components, the general quaternion being $t + xI + yJ + zK$, where the addition of the "temporal" or "scalar" part, t, corresponds to (5.2.2).

When I first came across the quaternionic expression of the rotation of a vector in 3D space, i.e., $\tilde{p} = qpq^{-1}$, I was puzzled as to why this expression is quadratic in the "rotation operator", q. This was because, in the more familiar 3D matrix notation, a rotation would be written as a real 3 x 3 rotation matrix acting on the vector of x, y, z components, $\bar{v}' = (R)\bar{v}$. For example, a rotation about the z-axis by an angle θ would be written,

$$\begin{pmatrix} v'_x \\ v'_y \\ v'_z \end{pmatrix} = \begin{pmatrix} cos\theta & -sin\theta & 0 \\ sin\theta & cos\theta & 0 \\ 0 & 0 & 1 \end{pmatrix} \begin{pmatrix} v_x \\ v_y \\ v_z \end{pmatrix} \qquad (5.4.2)$$

Hence, the relationship between the rotated and initial coordinates is linear in the components of the rotation matrix (R). Why is it, then, that the quaternion expression of a rotation is quadratic in the quaternionic rotation operator, q?

The reason is that, whilst quaternions were developed by Hamilton to describe rotations of 3D vectors, that is not their natural application. I do not know whether Hamilton himself appreciated this, but the fundamental object upon which quaternions act is not 3D vectors nor the 4D entities $t + xI + yJ + zK$ upon which Hamilton's attention largely focussed. Quaternions naturally operate upon an entity with two complex components, as their representation in terms of Pauli matrices betrays. I will refer to these entities as 2-spinors, a name which derives from their connection with angular momentum (though I'll not be justifying that connection here – for details see *The Unweirding* or any standard quantum mechanics text). In maths-speak, ordinary 3D vectors are actually second rank tensors in the spinor basis.

It is worth confirming that the quaternionic relationship $\tilde{p} = qpq^{-1}$ does work, for example by reproducing (5.4.2) for a rotation about the z-axis. In this case we have $q = exp\{\theta K/2\}$ and $q^{-1} = exp\{-\theta K/2\}$. Because $K^2 = -1$ it follows that the equivalent of de Moivre's Theorem applies and so $exp\{\theta K/2\} = \cos\frac{\theta}{2} + K \sin\frac{\theta}{2}$ and $exp\{-\theta K/2\} = \cos\frac{\theta}{2} - K \sin\frac{\theta}{2}$. Hence, (5.4.3)

$$\tilde{p} = qpq^{-1} = \left(\cos\frac{\theta}{2} + K \sin\frac{\theta}{2}\right)\left(v_xI + v_yJ + v_zK\right)\left(\cos\frac{\theta}{2} - K \sin\frac{\theta}{2}\right)$$

I'll leave the simplification of this expression as an exercise for the reader, hence showing that it reproduces (5.4.2). You will need to note that the quaternion algebra (5.4.1) leads immediately to $IJ = K = -JI$ and similar expressions, and $KIK = I, KJK = J$. We note in passing that these relations provide the commutation relations,

$$[I,J] = 2K \qquad [J,K] = 2I \qquad [K,I] = 2J \qquad (5.4.4)$$

which simply reproduce (5.2.6), but after allowing for the factors of i, that is $\sigma_x = iI$, etc. Whilst the Pauli matrices are Hermitian, the basis quaternions, I, J, K, are anti-Hermitian.

5.5 The Dirac Equation and Its Implications

In 1928, following the approach described above, Dirac was led to an equation in the complex-valued, 4-component entities, ψ, which I will refer to as 4-spinors, giving the famous Dirac equation,

$$(5.5.1)$$

$$E\psi = \left(\sqrt{p^2 + m^2}\right)\psi = (\alpha_i p_i + \beta m)\psi = \left[\begin{pmatrix} 0 & \sigma_i \\ \sigma_i & 0 \end{pmatrix} p_i + m \begin{pmatrix} \mathbb{I} & 0 \\ 0 & -\mathbb{I} \end{pmatrix}\right]\psi$$

where summation over the three Cartesian coordinates is implied by the repeated subscript i. Strictly, the Dirac equation proper would have the energy and momenta in (5.5.1) replaced by their differential operators, (5.1.2).

Recall that Dirac was trying to achieve two things: (i) a relativistic version of the Schrodinger equation, and, (ii) elimination of the negative energy solutions that were implied by (5.1.4). I'll address the first of those objectives shortly, but as for the second…there was an apparent disaster from which Dirac managed to snatch the most tremendous victory.

The disaster is that the Dirac equation, (5.5.1), has failed to rid us of the negative energy solutions. There is no need to look further than the case of a stationary particle with zero momentum, $p_i = 0$. In terms of two-component quantities, u and v, the Dirac equation then has solutions,

$$\psi = \begin{pmatrix} u \\ 0 \end{pmatrix} \qquad \text{and} \qquad \psi = \begin{pmatrix} 0 \\ v \end{pmatrix} \qquad (5.5.2)$$

The first of these solutions has energy $E = m$, i.e., the rest-mass energy, as we would expect for a stationary particle. But the second solution has negative energy, $E = -m$. So Dirac failed in his mission to get rid of the negative energy solutions.

Putting aside the negative energy solutions for a moment, Dirac's equation was successful in another way. Whilst Schrodinger's and Pauli's 1926 papers correctly derived the dependence of the spectral lines of the hydrogen atom on the principal quantum number, $n = 1,2,3$ …. they failed to explain the finer detail of the spectrum. Dirac's equation also correctly predicted the fine structure of the spectral lines due to the dependence of the electron energy levels on the angular momentum quantum number, j (the combined effect of electron spin and orbital angular momentum). Moreover, Dirac's equation automatically gives rise to the correct (first order) gyromagnetic ratio of the electron, namely $g = 2$, which was previously hard to understand (naïve theories apparently suggesting $g = 1$).

However, even as regards the awkward negative energy solutions, some progress has been made in that the negative energy solutions are distinguished from the positive energy solutions in terms of their differing contributions to the four components of the 4-spinor. One should not, however, make the mistake of thinking that the upper pair of components always correspond to positive energy, and the lower pair to negative energy, as is the case in (5.5.2). For moving particles, all four components of the 4-spinor will, in general, be non-zero for both positive and negative energy solutions. Nevertheless, the positive and negative energy solutions can always be distinguished by the nature of their spinor components. This is manifest by an orthogonality relationship which applies for any momentum. Hence, if $\psi_+(\bar{p})$ is the 4-spinor for a positive energy solution with momentum $\bar{p}$, and $\psi_-(\bar{p})$ is the 4-spinor for a negative energy solution with momentum $\bar{p}$, then $\bar{\psi}_+(\bar{p})\psi_-(\bar{p}) = 0$, where the conjugate spinor is defined by $\bar{\psi} = \psi^+\beta$. (See any standard relativistic quantum mechanics text for details, e.g., Bjorken and Drell).

Because the positive and negative energy solutions are distinguishable, the option is open to interpret them as representing different particles. This was to be the ultimate interpretation and the way out of the problem of negative energy.

Initially, Dirac turned failure into triumph in a slightly different way. To solve the problem of electrons cascading down to every lower negative energies he envisaged the vacuum to have all its negative energy electron states already filled. Pauli's Exclusion Principle then forbids any further electrons moving into negative energy states. In my opinion the infinite negative charge density that this idea imposes on the vacuum is rather a flaw in the conception, though Dirac argued, quite rightly, that the uniform nature of this charge distribution would render it unobservable. The fecundity of the idea, however, was that an electron might be raised out of its negative energy state, by supplying sufficient energy to do so, leaving a 'hole' in the otherwise uniform sea of negative charge. This hole would then appear to be a positively charged particle.

Dirac's initial idea (Dirac, 1930) was that this positive particle he had just postulated might be the proton. This was natural because the only known particles at the time were the electron and the proton. Even the neutron was not yet known, James Chadwick making the discovery only in 1932. While Rutherford had postulated the existence of such a neutral particle long before, this was not a generally accepted idea. The consensus at the time is proved by Gamow's 1931 monograph on the Constitution of Atomic Nuclei and Radioactivity, which made no mention of any neutral particle. Instead nuclei were supposed to consist of as many protons as necessary to get the mass about right, plus a sufficient number of electrons to also get the net charge right. The latter were known as "nuclear electrons". This is important context as it shows how immature was the prevailing state of knowledge, and hence makes even more remarkable Dirac's bold claim for anti-matter.

But Dirac's idea that his postulated "holes" were protons actually flounders immediately because the Dirac equation itself requires the positive and negative energy states to have the same mass (whereas, of course, the proton is 1836 times more massive than the electron). There were other fatal flaws in the idea too, such as rendering all atoms unstable. After stunning the world of physics with his relativistic equation in 1928, the period 1929 to 1930 saw Dirac being lauded in the press but his star rather in a decline with the physicists that counted. Landau reported back to Bohr regarding a lecture by Dirac he had attended with the single word "crap". The hole theory was not meeting with much approbation. Even his friend Heisenberg wrote, "The saddest chapter of modern physics is and remains the Dirac theory", Farmelo (2009). According to David Vidmar, that quote continued, "...I regard it as learned trash which no one can take seriously". Oops, bad call, Heisenberg. This was despite Heisenberg's earlier opinion that Dirac was so clever that no one could compete with him.

By 1931 Dirac had accepted that holes could not be protons. But, despite the widespread disbelief of colleagues, he was instead proposing that holes were a new type of particle with the same mass as the electron but opposite charge, which he was now calling an anti-electron. The tenaciousness of his belief that holes were physically real is remarkable given the opposition to the idea and the fragility of the theoretical case. One supposes that the need for his spinors to have (at least) four components, as we have seen, convinced him that the positive and negative energy solutions must be equally real. Sound reasoning, I think, but it is not easy to stick to one's guns with most of the senior physics establishment laughing at the idea. Nor did Dirac have the comfort of foreseeing any possibility of testing his prediction in the near future.

And yet in November of that same year, 1931, Carl Anderson, a student of Millikan's, had already pointed out to his mentor that his balloon-borne cloud chamber photographs showed "very frequent occurrence of simultaneous ejection of electron and positive particle". By December 1932 it was becoming clear that the positron had been found – many times over. And whilst the American, Anderson, is acknowledged as having precedence in the discovery, by the first months of 1933, Blackett and Occhialini at the Cavendish laboratory, Cambridge, also had many cloud chamber photographs showing particles with both charges and the same mass. By the end of 1933, even Dirac's detractors were grudgingly accepting that the positron existed. (It was Anderson, not Dirac, that dubbed the particle the "positron"). With the benefit of hindsight, the positron would have been easy to discover some years earlier if experimentalists had taken Dirac's prediction seriously. Paul Dirac and Carl Anderson won the Nobel Prize in 1933 and 1936 respectively.

5.6 Relativistic Covariance of the Dirac Equation

Whilst Dirac's first objective, to get rid of the negative energy solutions, did not work out quite as planned, his other objective, namely to derive a quantum equation which was fully relativistic, was achieved admirably. In a way this is not surprising because the equation is, after all, a "square root" of the relativistic equation for energy, $E^2 = p^2 + m^2$. However, that is not quite sufficient to prove the matter once the Dirac matrices and 4-spinors enter the picture. Relativistic covariance of the Dirac equation, (5.5.1), requires us to specify how the 4-spinors transform between observers in relative motion, and whose coordinate systems might be rotated, one with respect to the other. In (special) relativity, such a transformation between observers is a Lorentz transformation.

I gave some simple examples of a Lorentz transformation in §3.7. However, in that section the transformations were restricted to relative motion in the x-direction without rotation of the coordinates. In general, a Lorentz transformation can involve relative motion in an arbitrary direction together with an arbitrary 3D rotation. To begin with let's consider such an arbitrary Lorentz transformation. Before we start, let's put the Dirac equation in a more elegant form which shows space and time in a symmetrical manner. For this purpose, define Dirac's γ-matrices by,

$$\gamma_0 = \beta, \ \gamma_i = \beta\alpha_i \tag{5.6.1}$$

The Dirac equation, (5.5.1), becomes, after multiplying by β,

$$(\gamma_0 E - \gamma_i p_i - m)\psi = (\gamma_\mu p^\mu - m)\psi = 0 \qquad (5.6.2)$$

Here I have used relativistic notation for the contravariant components of the momentum 4-vector, where $\{p^\mu\} \equiv (p_0, -p_i)$. (The usual convention is that Latin indices run over the three spatial coordinates, whereas Greek indices run over both space and time coordinates, the latter being given by the subscript 0. Summation over repeated indices is implied).

To be successful as a relativistic equation, all observers related by a Lorentz transformation should find that "the same" equation holds. Here "the same" means once we account for how the transformation affects the momenta, p^μ, **and** how it affects the 4-spinor, ψ. Note that latter point carefully: the components of the 4-spinor are also changed by the transformation. In other words, the spinor is a type of geometrical object whose representation in terms of components will vary according to the spacetime coordinate system (i.e., according to the observer's state of motion and orientation of his coordinate system). In this respect the 4-spinor is analogous to, but very different from, a vector or 4-vector.

Suppose a contravariant 4-vector transforms according to $p'^\mu = \Lambda^\mu{}_\alpha p^\alpha$, where p'^μ are the components of 4-momenta seen by the second observer when the first observer sees 4-momenta p^α, and the matrix $\Lambda^\mu{}_\alpha$ implements the Lorentz transformation for 4-vectors. The spinor also transforms under the same Lorentz transformation as $\psi' = S\psi$, where S is a 4 x 4 matrix operating on the components of the 4-spinor, ψ. The Dirac equation as seen by the transformed observer is therefore,

$$(\gamma_\mu p'^\mu - m)\psi' = (\gamma_\mu \Lambda^\mu{}_\alpha p^\alpha - m)S\psi = 0$$

Multiplying by the inverse spinor operator, S^{-1}, gives,

$$(S^{-1}\gamma_\mu S\Lambda^\mu{}_\alpha p^\alpha - m)\psi = 0$$

Hence, this will always agree with the Dirac equation 'seen' by the other observer iff,

$$S^{-1}\gamma_\mu S\Lambda^\mu{}_\alpha \equiv \gamma_\alpha$$

If we multiply by the inverse of the matrix $\Lambda^\mu{}_\alpha$ we get,

$$S^{-1}\gamma_\mu S \equiv \Lambda_\mu{}^\alpha \gamma_\alpha \qquad (5.6.3)$$

(See Appendix J for explicit expressions for the matrices $\Lambda^\mu{}_\alpha$ and $\Lambda_\mu{}^\alpha$, though these are not needed for our purposes). The RHS of (5.6.3) transforms the entities γ_μ as if they were just 4-vectors, summing over the $\gamma_0, \gamma_x, \gamma_y, \gamma_z$ but without mixing their matrix components. The LHS of (5.6.3), however, does the opposite. For a given μ, only one matrix is involved, but the LHS acts on its components, summing over them. So (5.6.3) is a highly non-trivial requirement. It defines the transformation, S, which we must impose on the spinor, i.e., $\psi' = S\psi$, if the Dirac equation is to be truly relativistic.

The LHS of (5.6.3) might remind you of something? I refer to the way in which a position 'vector' expressed in the form of a quaternion, p, transforms, which we saw in §5.4 was $\tilde{p} = qpq^{-1}$. Here $\tilde{p}$ is the transformed position 'vector' in quaternion form, and q is the quaternion which implements the transformation. This is closely analogous to (5.6.3) with q being replaced by S^{-1}. It should not be too surprising that this turns out to be so, because the Dirac γ-matrices are related to the Pauli matrices, which are representations of the quaternions. This gives us a big clue as to the explicit expression for the 4-spinor transform, S. It is this,

$$S = S_{\mu\nu}(w) = exp\left\{-\frac{i}{4}w\sigma_{\alpha\beta}\varepsilon_{\mu\nu}^{\alpha\beta}\right\} \tag{5.6.4}$$

where the subscripts $\mu\nu$ label the axis of the Lorentz transformation, w is its magnitude, and $\sigma_{\alpha\beta} = \frac{i}{2}\left[\gamma_\alpha, \gamma_\beta\right]$ and $\varepsilon_{\mu\nu}^{\alpha\beta} = \delta_{\alpha\mu}\delta_{\beta\nu} - \delta_{\alpha\nu}\delta_{\beta\mu}$ are a set of antisymmetric 4 x 4 matrices which depend upon whether a rotation or a 'boost' (change of velocity) is considered, and about what axis. For example, $\varepsilon_{0x}^{\alpha\beta}$ is a boost is the x-direction, of magnitude given by the "boost parameter", w (see Appendix K). The only non-zero components of $\varepsilon_{0x}^{\alpha\beta}$ are $\varepsilon_{0x}^{01} = 1$ and $\varepsilon_{0x}^{10} = -1$. See Appendix K for how $S^{-1}\gamma_\mu S$ reproduces the correct Lorentz boost of a 4-vector.

Alternatively, $\varepsilon_{xy}^{\alpha\beta}$ represents a rotation about the z-axis by angle w. The only non-zero components of $\varepsilon_{xy}^{\alpha\beta}$ are $\varepsilon_{xy}^{12} = 1$ and $\varepsilon_{xy}^{21} = -1$. In the next section we shall see how this correctly reproduces the rotation of the spatial part of a 4-vector.

Just as quaternions can be used to represent both a position vector (p) and also a rotation operator (q) in 3D space, so (5.6.3) indicates how the Dirac matrices can be used to define both a 4-vector and Lorentz transformations of the 4-vector. To this end, the γ_μ matrices act like unit vectors in spacetime (analogous to the base quaternions I, J, K), so the 4-vector with contravariant components a^μ can be written $a^\mu \gamma_\mu$ (summation assumed). A Lorentz transformation of the 4-vector can now be implemented by acting on the "unit vectors", operating on the matrix components of the γ_μ matrices according to (5.6.3). Hence,

$$S^{-1} a^\mu \gamma_\mu S \equiv a^\mu \Lambda_\mu{}^\alpha \gamma_\alpha = a'^\alpha \gamma_\alpha \tag{5.6.5}$$

where $a'^\alpha = a^\mu \Lambda_\mu{}^\alpha$ is the Lorentz transformation of the 4-vector components. There is a subtlety lurking in (5.6.5), however, as the transformation $a'^\alpha = a^\mu \Lambda_\mu{}^\alpha$ is the inverse of the usual $a'^\mu = \Lambda^\mu{}_\alpha a^\alpha$. This is as it should be because one may rotate a vector wrt fixed axes, or rotate the axes wrt a fixed vector. The effect on the coordinates is the same provided the two rotations (or boosts) are in opposite senses. This is spelt out in detail in Appendix J.

5.7 That Pesky Minus Sign

The success of Dirac's programme in revealing the existence of antimatter is only one part of what he accomplished. Another aspect of the Dirac equation is its invariance under arbitrary Lorentz transformations. Let us leave the relativistic part alone. ignoring boosts to different speeds. Let us concentrate more simply on just rotations, which are also Lorentz transformations. Based on (5.6.4), Appendix K has derived the spinor transformation matrix, S, which tells us how to find the components of the rotated 4-spinor, ψ', in terms of the original 4-spinor, ψ, namely by using $\psi' = S\psi$, for the specific case of a rotation about the z-axis by an angle θ,

$$S(\theta) = \mathbb{I}\cos\left(\frac{\theta}{2}\right) - i\sigma_{xy}\sin\left(\frac{\theta}{2}\right) \tag{5.7.1}$$

where $\mathbb{I}$ is the 4 x 4 unit matrix and $\sigma_{xy} = \begin{pmatrix} \sigma_z & 0 \\ 0 & \sigma_z \end{pmatrix}$. Note that if the 4-spinor is written as a pair of two-component spinors, $\psi = \begin{pmatrix} u \\ v \end{pmatrix}$ then both u and v are transformed independently and identically under the rotation, i.e., $u' = \tilde{S}u$ and $v' = \tilde{S}v$, where,

$$\tilde{S}(\theta) = \mathbb{I}\cos\left(\frac{\theta}{2}\right) - i\sigma_z\sin\left(\frac{\theta}{2}\right) \tag{5.7.2}$$

where $\mathbb{I}$ is now the 2 x 2 unit matrix.

Unlike the rotation of a vector, e.g. (5.4.2), the transformation operators in (5.7.1) or (5.7.2) do not depend upon $cos(\theta)$ and $sin(\theta)$ but upon $cos\left(\frac{\theta}{2}\right)$ and $sin\left(\frac{\theta}{2}\right)$. The implications are profound. A rotation through a full revolution, that is through an angle $\theta = 2\pi$, leaves any vector unchanged, of course. That is manifest by (5.4.2) becoming the unit matrix, because $cos(2\pi) = 1$ and $sin(2\pi) = 0$. But inserting $\theta = 2\pi$ into (5.7.1) gives $S = -\mathbb{I}$. So a 4-spinor is a very peculiar object which does not return to its original value after rotation through a full revolution but instead reverses in sign. The same is true of (5.7.2) for the 2-spinors, so this acquisition of a minus sign on full rotation is not due to a relativistic effect: it occurs for 2-spinors also, which have no relativistic content. This is found for a rotation by 2π about any axis. It requires two full revolutions, in other words a rotation through an angle of $\theta = 4\pi$, to return a 4-spinor or a 2-spinor to its original value.

Note that, because (5.6.5) is quadratic in S, the minus sign cancels, so (5.6.5) is consistent with a 4-vector being unchanged by a single full rotation, i.e., by 2π. So we see, once again, that the minus sign that applies to S or $\tilde{S}$ after a rotation by 2π arises, in some vague sense, from this spinor transform, S, being the "square root" of the 4-vector transform Λ. In other words S is quadratically related to Λ in (5.6.5).

The implication is that, for spin ½ particles, i.e., particles whose behaviour is governed by the Dirac equation, rotating through a full revolution will reverse the sign of their wavefunction (which, in this context, is just another name for the 4-spinor, ψ). The physical implication of this is that, in any interference experiment involving rotating particles of spin ½, destructive interference will occur after a 2π rotation. This is observed in such experiments, e.g., in neutron diffraction which is commonly used as a tool in the material sciences.

More generally, for any rotation matrix, Λ, acting on a 4-vector, there are two different 4-spinor operators which reproduce that same Λ via (5.6.5), namely both $S(\theta)$ and $S(\theta + 2\pi) = -S(\theta)$. In slightly imprecise maths-speak, the 4-spinors are said to provide a double cover of the 4-vectors.

[Optional aside: The more precise maths-speak is as follows. Confine attention to ordinary 3D vectors, or 3-vectors, which transform under rotations according to orthogonal matrices with unit determinant. These matrices are (or, rather, form a representation of) the group SO(3), which stands for Special Orthogonal matrices of dimension 3 x 3, and "special" just means having determinant of 1. Consider now some two-component column matrices with complex components, in other words 2-spinors. If they are taken to transform under rotation according to the 2 x 2 complex-valued matrices like 5.7.2, and more generally matrices of the form $q = exp\left\{-\frac{i}{2}\theta\hat{n}\cdot\bar{\sigma}\right\}$, then these complex two-spinors stand in 2-to-1 correspondence to the real 3-vectors, the relationship being essentially the quaternionic version of 3D rotations. But these q matrices are the most general 2 x 2 unitary matrices (i.e., matrices whose complex transpose equals their inverse) which also have unit determinant, and these form the group SU(2), the Special Unitary group of 2 x 2 complex matrices. So, finally, the full maths-speak is that SU(2) is the double cover of SO(3)].

5.8 Is Matter the Square Root of Geometry?

I emphasise that the importance of the observations in §5.7 are not merely mathematical. Dirac's 4-spinors are a successful ***physical*** description of spin ½ particles. More simply, the 2-spinors are a successful description of spin ½ particles as regards their purely geometrical behaviour, that is their behaviour when their relationship with space is changed - specifically their behaviour under rotations. So the minus sign acquired by the wavefunction of electrons, protons and neutrons (and muons and tau particles and quarks and neutrinos) upon rotation by a full 2π has physical implications, as already noted. Neutron interference effects are perhaps the best experimental confirmation, see Rauch and Werner (2018) or similar sources.

Let me emphasise this some more. ***All*** the "matter" type particles of the standard model of particle physics are of spin ½. The other particle types in the model are all "force carriers" and of spin 1 (the photon, the gluons and the W and Z particles), or the spin 0 Higgs particle. The key difference between the "matter type" particles with spin ½ and the "force carriers" with integer spin is that Pauli's exclusion principle applies to the former but not to the latter. Multiple particles with integer spin (bosons) can occupy the same quantum state. Hence, macroscopically large numbers of such quanta can pile up to create a classical field (e.g., the electromagnetic field).

In contrast, no two identical fermions (half-integer spin particles) can occupy the same quantum state. As a result, these "matter particles" retain their separate particle-like status even when there are many of them. Moreover, the exclusion principle is essential in preventing all atomic electrons dropping down into the ground S-state and rendering all chemistry void. Moreover, the exclusion principle turns out to be essential to the stability of matter at all. Any mixture of positive and negative charges would be unstable and collapse in a heap if the particles were all bosons (integer spin). It is essential that at least one sign of charge consists of fermions (half-integer spin). The proof that matter can then be stable is notoriously difficult and was first proved by Dyson and Lenard in 1967. A more accessible proof was provided subsequently by Lieb and Thirring (1975).

And so, in conclusion,

- in as far as particles may be identified with their quantum states,

- and in as far as the quantum states of spin ½ particles are 4-spinors (and hence behave like 2-spinors under rotations),

- and in as far as the spatial behaviour of spinors is related to conventional geometry as a type of "square root", requiring two factors of S to make one Λ, see (5.6.5) or the generalised quaternion version of Appendix L, $p' = qpq^{*\dagger}$, and other variants within Appendix L,

- and in as far as "matter" can be identified with spin ½ particles,

- it follows that matter is a type of square root of geometry.

5.9 Dirac's Belt Trick

It is not just exotica like elementary particles which have the peculiar property of (i) changing upon rotation by 2π, and yet, (ii) being unchanged by rotation by 4π. There are many variants on the theme of simple examples of such behaviours, but my favourite is Dirac's Belt Trick. If you have never done this for yourself, I urge you to do so now. If your intuition is like mine, it will refuse the fence. You see it, but you cannot grasp what's happening.

Before trying the belt trick yourself, you might benefit from watching a video of it done, such as the YouTube video by NoahExplainsPhysics (see References). You can find other videos of the belt trick on the internet; here's mine: http://RickBradford.co.uk/DiracBeltTrick.avi.

But I would still encourage you to do the belt trick yourself. These are the step by step instructions,

- Take any long, flat, flexible object – a belt is ideal;

- Secure one end of the belt on a desk top, e.g., by placing a heavy object on it so it cannot move;

- Hold the other end of the belt in your hand and stretch the belt out straight, without any twists;

- Then rotate your end of the belt through a full 360° rotation;

- The belt is now twisted, obviously. If you hold your end of the belt so that it cannot rotate, you will find that however you move the end around it will not result in an untwisted belt once you pull it straight again (though you may find you can reverse the sense of the twist!). You will hardly be surprised. But…

- Starting with the belt straight and untwisted again, this time rotate your end through two full rotations (720°);

- On the face of it, the belt is now twice as twisted as before. So if you could not undo the twist before, surely you won't be able to now? Wrong;

- Holding the end so it cannot rotate, move the end around the back of the belt and then pull it straight again. The twist will have vanished!

That's Dirac's belt trick.

As I said, my intuition refused to assimilate how this works. But the rest of this section explains it. It will explain why the twist due to a single rotation cannot be undone, but the twist due to a double rotation can. The secret is essentially topological.

Put all thought of spinors out of your mind. We are now considering ordinary everyday geometry. First let me show what part non-trivial topology plays in rotations in 3D space. In 3D space we can describe an arbitrary rotation by a unit vector, $\hat{n}$, defining the axis of rotation, and by the angle of rotation, θ. But adding 2π to θ is just the same rotation. To express that differently, a rotation by π is the same thing as a rotation by $-\pi$ about the same axis. We may thus consider a space which defines uniquely all possible rotations in 3D by a three dimensional ball of radius π. The vector $\theta\hat{n}$ drawn from the origin therefore uniquely defines a 3D rotation, i.e., there is a one-to-one correspondence (bijective map) between points in this ball of radius π and 3D rotations. BUT that claim is only correct if the ball has diametrically opposite points identified, because a rotation by π is the same thing as a rotation by $-\pi$ about the same axis. That requirement is what renders the topology of this "space of rotations" non-trivial.

Now suppose you consider a rotation about, say, the axis $\hat{n} = (1,1,0)/\sqrt{2}$. Suppose the angle of rotation gradually increases from 0 to, say, 270°, that is $3\pi/2$ radians. Figure 5.1 illustrates the path taken in our "space of rotations" as the rotation angle increase from zero.

Figure 5.1: Rotation about axis $\hat{n} = (1,1,0)/\sqrt{2}$ increasing from zero to 270°

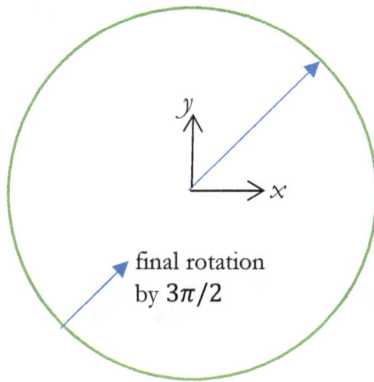

(The diagram should show a 3D ball really, but I've placed the rotation axis in the x, y plane so I could get away with a two dimensional sketch. Please mentally insert the third dimension).

Figure 5.1 illustrates how crucial is the identification of diametrically opposite points. Without that, how would we represent a sequence of rotations from zero to $3\pi/2$? We would get a straight radial line of length $3\pi/2$ (and hence outside the "sphere of rotations") and this would give no indication that it was, in fact, equivalent to a rotation of $-\pi/2$. By identifying diametrically opposite points, we re-enter the ball at the opposing point when the rotation exceeds π.

This may seem rather trivial. But it is not. It is the key to understanding the belt trick.

The first step is to free your mind from thinking that the only rotation of relevance is the rotation you applied to the end of the belt. The rotation at the fixed end of the belt is zero, for a start. Imagine a sequence of coordinate systems embedded in the belt – say with the x-axis along the belt length, the y-axis across its width, and the z-axis perpendicular to the belt. With the belt pulled straight, the rotation you apply to the end is about the x-axis.

But suppose you do not apply any rotation to the end of the belt. Instead you move your hand holding the end of the belt to one side so that the belt is curved. This will unavoidably cause the coordinate systems embedded in the belt to be rotated, but this rotation will be a continuous function of position along the belt. However, the rotation has been constrained to be zero at both ends of the belt. If we plot the locus of points which represent the rotations at different points along the belt on our spherical "space of rotations" we get something like Figure 5.2. The locus necessarily starts and finishes at the origin (zero rotation), and so is a closed loop of some sort.

Figure 5.2: The belt with no rotation at both ends, but displaced from straight so as to induce rotations along the length of the belt. *(taken from NoahExplainsPhysics)*

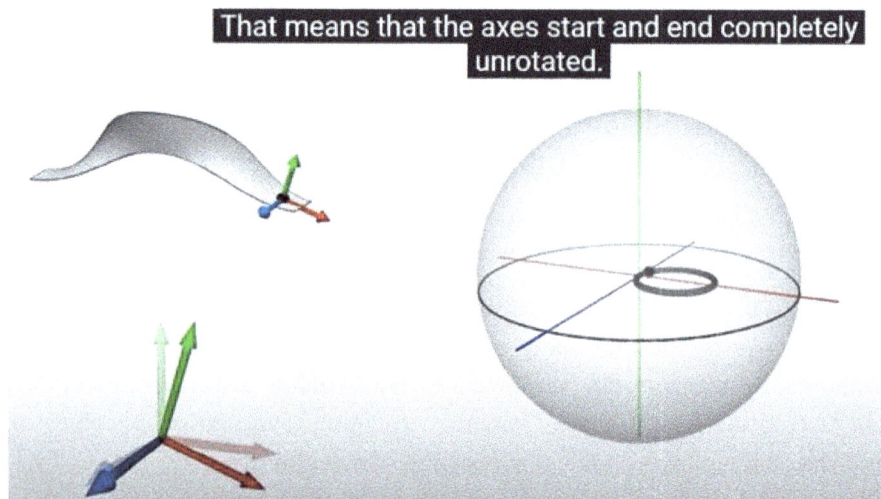

That means that the axes start and end completely unrotated.

If we now slowly pull the belt straight again, so as to remove the twist we have induced in it, the loop shrinks and becomes a single point at the origin as the belt becomes straight and untwisted again.

More generally, any configuration of the belt can be represented by a locus in our "space of rotations". If the rotation at both ends of the belt are constrained to be zero, then the locus representing the belt is a closed loop starting and finishing at the origin. Our ability to remove the twist in the belt can be identified with our ability to **continuously** deform such a closed loop to a point at the origin.

That is a topological property of trajectories in our spherical "space of rotations" (specifically, its homotopy) – and recall that this space has a non-trivial topology. You may sense where this is leading.

Suppose, with the belt initially straight and untwisted, you apply to the end a full 2π rotation. The rotation at each point along the belt maps to a point in our spherical "space of rotations". The locus of such points representing this initial configuration of the belt, which is still straight but now twisted through a full rotation, lies along the x-axis, like Figure 5.3.

Figure 5.3: The representation of the whole belt by a locus of points in the "space of rotations": the initial situation with the belt straight and a rotation of 2π applied at the end.

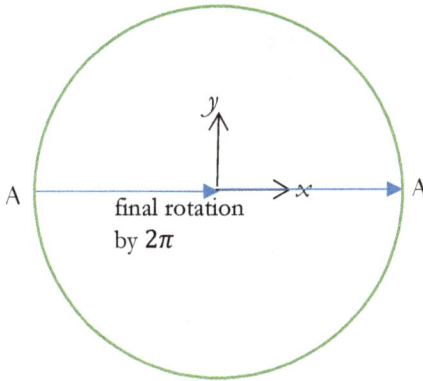

Though it does not look like it, the straight line locus in Figure 5.3 is also topologically a closed loop, because points A, on both ends of the diameter, are the same point. But this is now a closed loop of a very different topological kind from that of Figure 5.2, and in a physically crucial respect. This closed loop cannot be continuously shrunk to the origin, and so we cannot get rid of the twist in the belt by moving around the end whilst not allowing the end to rotate. To see why the locus cannot be shrunk to a point, try deforming it in an arbitrary way – but always respecting the identification of diametrically opposite points, and keeping the ends form rotating (and so at the origin in Figure 5.3). You can only ever get something like Figure 5.4 (though you have to imagine this in a 3D sphere, the implication is the same).

What we are dealing with here is a topological impossibility.

Figure 5.4: The impossibility of shrinking to a point a closed locus that winds around sphere.

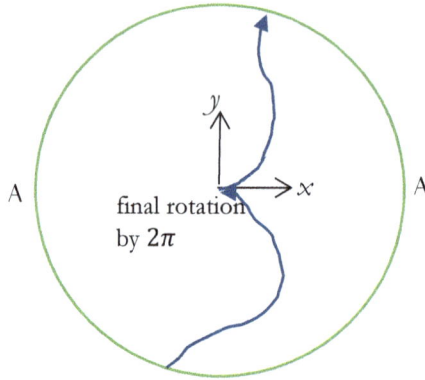

So far, so good. But you probably didn't need this degree of sophistication to convince you that a 2π twist in the belt could not be removed just by moving the end around without rotating it. You are hardly surprised.

The real point of all this development is to explain what happens when the initial rotation applied to the belt is 4π. Figure 5.3 shows the locus for a 2π rotation. To get a 4π rotation we need to go around the "sphere of rotations" again. Figure 5.5 shows the initial configuration of the straight belt after a 4π rotation of its end. The rotations in Figure 5.5 are slightly displaced from being perfectly about the x-axis to show the two passes across the sphere more clearly.

The key issue is this: if we can continuously deform the two-pass locus in Figure 5.5 to a point at the origin, then we can untwist the belt without rotating the ends.

And we can.

Figure 5.5: The representation of the whole belt by a locus of points in the "space of rotations": the initial situation with the belt straight and a rotation of 4π applied at the end.

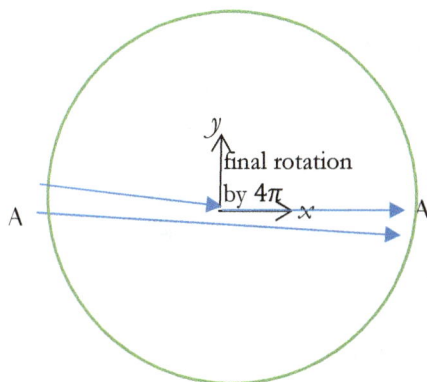

This has become possible precisely because there are now two passes across the sphere. NoahExplainsPhysics shows this very nicely in the ideal way, as a video. But hopefully the reader will be convinced by the sequence of deformations of the locus shown in Figure 5.6. Remember, the rule is that we must re-enter the sphere diametrically opposite where we leave it, and the locus starts and finishes at the origin.

Figure 5.6: The locus for a 4π rotation can be continuously deformed to a point at the origin. *All loci start along the black path. The complete locus is initially as given by Figure 5.5. It is first deformed to the blue locus, then to the purple locus, then the red locus. As the red locus shrinks onto point B, the left-hand part of the locus vanishes and we are left with a locus consisting only of the black line (from the origin to point B) followed by some return path from B to the origin, e.g., the green dashed curve. This closed loop can now be shrunk to the origin. QED.*

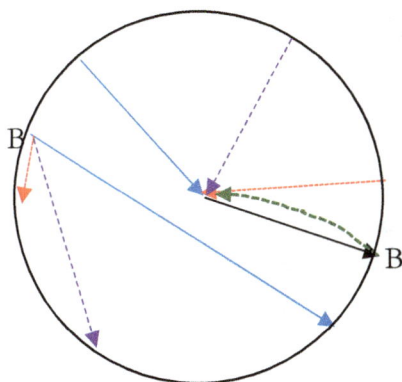

Figures 5.4 and 5.6 explain the belt trick, the former shows why a 2π twist cannot be undone, whilst the latter shows why a 4π twist can be undone.

5.10 The Double Covering Revisited

Recall in §5.7 we found that there were two different 2-spinor rotation operators (or, equivalently, two different quaternions) for every ordinary 3D rotation operator (i.e., every real 3 x 3 orthogonal matrix with unit determinant). The two different 2-spinor operators differ in sign, just as one would expect of the two square roots of the same item. So there are twice as many 2-spinor operators, manifest as 2 x 2 complex unitary matrices of unit determinant, as special 3 x 3 orthogonal matrices. They are in 2:1 correspondence. The maths-speak expression of this is that the group SU(2) is the double cover of the group SO(3).

Figure 5.7: The "space of rotations" for the double covering group of 2-spinors, SU(2). *The circles represent spheres. Points on their spherical surfaces are identified as illustrated by points A and B. Both spheres are of radius π. Point O is no rotation. Corresponding points in the two spheres represent the same real 3 x 3 rotation (element of SO(3)) but represent different 2-spinor rotation operators (elements of SU(2)). Point C is a rotation by 2π, hence has the same SO(3) element as point O (the unit matrix) but changes the sign of a 2-spinor (or quaternion). The arrows denote the increasing angle of rotation. Both 3-vectors and 2-spinors return to their initial value after return to point O (a rotation by 4π).*

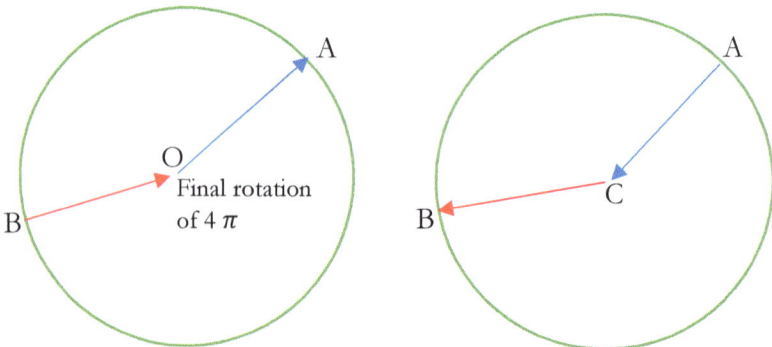

But if there are two 2-spinor rotation operators, i.e., elements of SU(2), for every real 3 x 3 rotation matrix (element of SO(3)), then how can we understand this in terms of our spherical "space of rotations"? Each point in that space corresponds to an element of SO(3), and hence to two different elements of SU(2). Hence, to create a new "space of rotations" which has a unique, one-to-one (bijective) mapping between its points and elements of SU(2) we need two spheres. These two spheres must be topologically connected in the manner shown by Figure 5.7.

Do note the analogy with Riemann surfaces. We saw in §2.17 how the idea of Riemann surfaces can make multi-valued functions single valued again. So, the square root function (acting on complex numbers) requires two Riemann surfaces with a non-trivial topological connection, thus separating the two square roots of different sign. This is closely analogous to Figure 5.7, though the topology is different because, in this case, the two joined spaces are 3D not 2D. Nevertheless, this further emphasises the analogy with square roots.

Figure 5.7 makes it easier to see why a 2π twist in the belt cannot be untwisted if its locus is fixed at points O and C. It is also easier to see how the locus for a 4π twist, that starts and finishes at point O, can be continuously deformed to a point at O. To do so we swing point B around to meet point A, in which case the locus need never enter the right-hand sphere at all.

So we see that, rather oddly, the belt is better represented by the double cover of SO(3) – the space of 2-spinor rotations. This is, of course, the reason why Dirac came up with the illustration.

Why should a very prosaic, everyday object – a belt – behave under spatial transformations akin to an object whose expression is crucially dependent upon complex numbers? Moreover, these latter objects, spin ½ particles, give us cause for a great deal of head-scratching over their physical properties. This is a good question. The answer would appear to be that feature of the belt which is initially overlooked: it is an extended object. The belt is not simply subject to a single rotation but defines a continuous locus of rotations – a locus within operator space.

Does this suggest that the true nature of spin ½ particles is that they are extended objects, in some manner like a belt? Who knows? But the whole business is certainly intriguing.

5.11 Quaternions and Physics

Because the pure quaternion algebra is isomorphic with that of the Pauli matrices, it is inevitable that we have bumped into them in this chapter. The full quaternion algebra, though, consists of the four component entities $t + xI + yJ + zK$. As the pure quaternion part, $xI + yJ + zK$, is taken to represent a vector in 3D space, Hamilton was led to consider the real part, t, to be time. Thus, Hamilton considered a quaternion to be a rather mystical marriage of space and time. This may be rather old hat now, but recall that this was 62 years before Einstein's 1905 relativity revolution. It has always seemed beguiling that a purely mathematical creation – inspired by the desire to represent 3D spatial rotations elegantly – necessitated the inclusion of a fourth coordinate. It does seem to point to greater revelations to come. And yet, despite some further intriguing results, quaternions have never quite lived up to their early promise.

The pure quaternions mimic vector algebra in an elegant fashion. If the real 3D vector $\bar{p}$ is represented by the quaternion $p_x I + p_y J + p_z K$ and similarly the real 3D vector $\bar{q}$ is represented by $q_x I + q_y J + q_z K$, then their quaternion product becomes, (using $IJ = -JI = K$, etc),

$$qp = -(q_x p_x + q_y p_y + q_z p_z) + (q_y p_z - q_z p_y)I + (q_z p_x - q_x p_z)J \\ + (q_x p_y - q_y p_x)K$$

This shows that the algebra of purely "vectorial" quaternions (i.e., quaternions with no scalar part) is not closed, as the above product includes a scalar part. Denoting the scalar and vector parts by $\mathbb{S}$ and $\mathbb{V}$ we see that, in terms of the usual scalar and vector products of 3D vectors,

$$\mathbb{S}(qp) = \mathbb{S}(pq) = -\bar{p} \cdot \bar{q} \qquad \mathbb{V}(qp) = -\mathbb{V}(pq) = \bar{q} \times \bar{p} \qquad (5.11.1)$$

This is pretty neat and so it is not surprising that James Clerk Maxwell ultimately adopted quaternions as his chosen vehicle for expressing his famous equations of electromagnetism. To be historically accurate, in Maxwell's early work on electromagnetism, namely the sequence of publications from 1855 to 1865, he gave the equations in component form, rather than any more compact notation. But with 1873's publication of the first edition of "A Treatise on Electricity and Magnetism" he deployed quaternions. The full timeline of the changing notations used for Maxwell's equations has been given by Hehl (2010).

Despite Hamilton's proselytising of quaternions, their use gave way to vector algebra under the influence of Grassmann, Gibbs, Heaviside and Lord Kelvin. The latter was the leading physicist of the mid- to late nineteenth century and was reputed to have a dim view of quaternions. Kelvin is quoted as opining, "Quaternions were invented by Hamilton after his truly important work had been completed. Although beautiful and of ingenious origin, they have been a curse on anyone who has come into contact with them in any way". Later editions of "A Treatise on Electricity and Magnetism" published after Maxwell's death had the quaternions removed and replaced by vector analysis. (Kelvin himself avoided both quaternions and vectors, so might have been something of an old fuddy-duddy). From Einstein's general relativity (1916) onwards, the use of tensors became *de rigueur*, and ultimately the language of exterior calculus would take over.

These days, people for whom the type of approach exemplified by (5.11.1) appeals, specifically the uniting of the scalar and the vector products into the same entity, are more likely to adopt Clifford algebra or Geometrical Algebra, see for example Hestenes and Sobczyk (1984).

From the inclusion of time in the full quaternion, $q = t + xI + yJ + zK$, one might hope that (special) relativity, that is the Lorentz transform, would arise naturally. Lamentably, this is not so. The square modulus of the quaternion representing an 'event' (i.e., a point in spacetime) is Euclidean, not Minkowskian. That is,

$$q^\dagger q = (t - xI - yJ - zK)(t + xI + yJ + zK) = t^2 + x^2 + y^2 + z^2$$

where the dagger superscript denotes the quaternion-conjugate. In contrast, what we want to be invariant (thus defining the Lorentz transformations) is the indefinite Minkowski metric, $t^2 - x^2 - y^2 - z^2$.

This is terribly disappointing. This flaw can be patched up by moving to biquaternions, defined as adopting complex values for the components t, x, y, z. A spacetime event can then be denoted $q = t + ixI + iyJ + izK$ for real t, x, y, z. Then we do indeed get,

$$q^\dagger q = (t - ixI - iyJ - izK)(t + ixI + iyJ + izK) = t^2 - x^2 - y^2 - z^2$$

as desired (where the quaternion-conjugate does not include the complex conjugate). The use of biquaternions does then lead to a very compact description of the general Lorentz transform, which looks just like a rotation in quaternion form, and also provides a single-equation description of the whole set of Maxwell equations. I show how these are derived in Appendix L, following the lovely review by Joachim Lambek (2013).

Quaternions have always been rather marmite to physicists and mathematicians. Minkowski abhorred them, but Dirac himself was reputed to be keen to give quaternions a greater role in physics. At the same time Dirac was dead against the use of biquaternions (see Lambek's review for a personal anecdote). I can understand how he felt. After the disappointment that the natural metric of quaternions is not Minkowskian, one feels so let down that the adoption of biquaternions seems ugly and artificial. However, that position presents the adherent of quaternions with an impasse.

Lambek's review shows how to shoe-horn the Dirac equation into a sort-of quaternionic form, but it is rather contrived and seems to offer nothing, neither in terms of notational compactness nor as regards physical insight. It is probably more enlightening to see the origin of the four component spinors as being the group representation theory of the Poincaré group. Here we find that the smallest dimension irreducible representation is the direct sum of two 2-spinor representations of SU(2), which in standard notation is written $\left(\frac{1}{2}, 0\right) \oplus \left(0, \frac{1}{2}\right)$. This has the advantage of being rather more transparently interpretable as two spin ½ particles, which are "joined at the hip" by virtue of the symmetries of spacetime.

Dirac was, incidentally, passionately opposed to quantum field theory, despite having been one of its initial creators. So Dirac ultimately joined the long list of early proponents of quantum mechanics who later had profound misgivings about its later developments or its interpretation.

5.12 Quaternions and Analysis

Because the use of complex numbers imbues much of mathematical analysis with its power, the discovery of quaternions naturally leads to an expectation of even greater riches of analytical power to be had through their adoption. Lamentably, they have failed thus far to live up to that expectation. However, one manifestation of the analytical power of the complex number field is the Cauchy integral theorem, (2.8.3). There is a quaternion equivalent which I derived myself many years ago. I will simply state it, but you can find the proof on my web site. I define a function, f, of a quaternionic variable, $q = t + ix + jy + kz$ to be q-holomorphic within a given region, δV^4, if $D^* f = 0$ everywhere in the region, and where, as in Appendix L, $D = \partial_t - i\overline{\nabla}$. If S_q is the closed, simply connected 3-surface which is the boundary of δV^4 then,

$$\oiiint_{\delta V^4} d^3 S_q f(t, x, y, z) = 0 \qquad (5.12.1)$$

This generalises the holomorphic integral theorem, §2.7. A generalisation of the Cauchy integral theorem, (2.8.3), is then, (5.12.2)

$$\oiiint_{\delta V^4} d^3 S_q H(t - t_0, x - x_0, y - y_0 z - z_0). f(t, x, y, z)$$
$$= \frac{2\pi^2 I}{3} f(t_0, x_0, y_0, z_0)$$

where H is the quaternionic function,

$$H(t, x, y, z) = e^{-t\bar{v}} \left(\frac{x}{r^4} \right) \tag{5.12.3}$$

which takes the place of $1/(z - z_0)$ in the Cauchy integral theorem. By symmetry, (5.12.2) also holds if I is replaced by J or K if the x in (5.12.3) is also replaced by y or z. The magnitude of the numerical factor on the RHS of (5.12.2) is simply the volume of a hypersphere, divided by 3 (equal shares for each spatial coordinate).

5.13 References

Bjorken, J.D. and Drell, S.D. *Relativistic Quantum Mechanics*. McGraw-Hill, 1964.

Dirac, P.A.M. (1928). *The Quantum Theory of the Electron*. Proceedings of the Royal Society of London. Series A. Vol.117, No.778 (Feb. 1, 1928), pp. 610-624. https://royalsocietypublishing.org/doi/10.1098/rspa.1928.0023

Dirac, P.A.M. (1930). *A theory of electrons and protons*. Proceedings of the Royal Society of London. Series A. Vol.126, No.801 (January 1930). https://doi.org/10.1098/rspa.1930.0013

Dirac, P.A.M. (1932). *Relativistic quantum mechanics*. Proceedings of the Royal Society of London. Series A. Vol.136, No.829 (02 May 1932). https://doi.org/10.1098/rspa.1932.0094

Farmelo, G. (2009). *The Strangest Man: the Hidden Life of Paul Dirac, Mystic of the Atom*. New York: Basic, 2009.

Hehl, F.W. (2010). *On the changing form of Maxwell's equations during the last 150 years*. UCL Department of Mathematics, Applied Mathematics Seminar, 22 February 2010. https://www.thp.uni-koeln.de/gravitation/mitarbeiter/hehl/MaxwellUCL2.pdf

Hestenes, D. and Sobczyk, G. (1984). *Clifford Algebra to Geometric Calculus: A Unified Language for Mathematics and Physics*. Springer, 1984.

Lambek, J. (2013). *In Praise of Quaternions*. CR Math. Rep. Acad. Sci. Canada, 2013, math.mcgill.ca.
https://www.math.mcgill.ca/barr/lambek/pdffiles/Quater2013.pdf

Lieb, .H. and Thirring, W.E. (1975). *Bound for the Kinetic Energy of Fermions Which Proves the Stability of Matter*. Physical Review Letters. 35 (11): 687–689.
http://doi.org/10.1103/PhysRevLett.35.687

NoahExplainsPhysics (23 August 2021). Dirac's belt trick, Topology, and Spin ½ particles. https://www.youtube.com/watch?v=ACZC_XEyg9U

Rauch, H. and Werner, S.A. (2018). *Neutron Interferometry: Lessons in Experimental Quantum Mechanics, Wave-Particle Duality, and Entanglement*. Oxford University Press, 2nd ed, 2018.

Vidmar, D. (undated). *The Dirac Equation and the Prediction of Antimatter*. dirac antimatter paper.pdf (ufrgs.br)

Appendix A: Proofs of Pythagoras's Theorem

Since Pythagoras features so centrally in chapter 1, and because it is so closely related to the discovery of the irrationals, I think I should include something on his famous Theorem: the square of the hypotenuse equals the sum of the squares of the other two sides. Anyone interested in maths is likely to have seen a proof along the lines of this one (though there are many variants),

Figure A.1: Proof of Pythagoras's Theorem

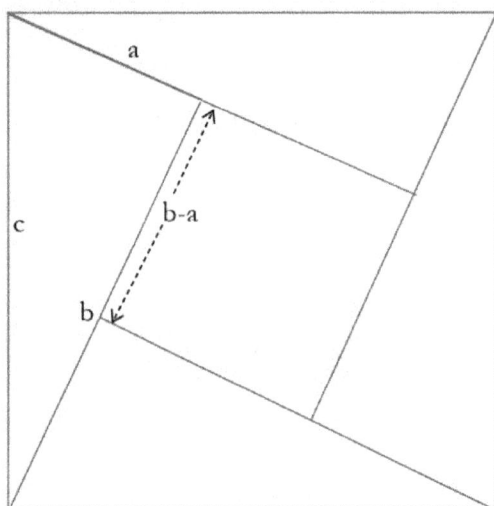

The outer square is of side c and each of the right-angled triangles have sides a and b where $a < b$. The smaller, inner square therefore has side $b - a$. The area of the outer square is thus c^2 and this must equal the sum of the areas of the four right-triangles plus the area of the inner square. The area of a right triangle is half the product of its base and height (i.e., $ab/2$) and the area of the inner square is $(b - a)^2$. So,

$$c^2 = 4 \times \frac{ab}{2} + (b - a)^2 = 2ab + (a^2 + b^2 - 2ab) = a^2 + b^2$$

QED.

Pythagoras's theorem relies upon the assumption that the geometry is Euclidean, which is manifest in this proof by the implicit assumptions regarding the formulae for the areas of a square and a right-triangle.

An alternative proof which does not use areas, but instead uses similar triangles follows. This method was reputedly used by Einstein as a boy.

Figure A.2: Another proof of Pythagoras's Theorem

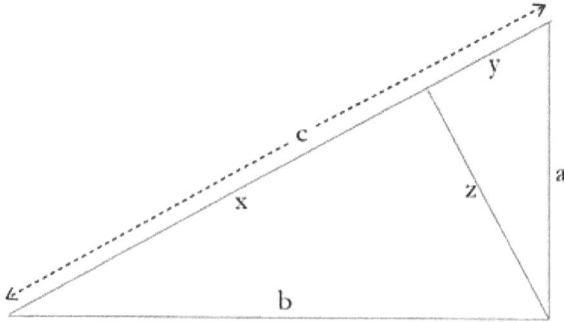

The three triangles all have the same angles, i.e., they are similar (same shape, different sizes). Also $c = x + y$. So,

$$\frac{b}{c} = \frac{x}{b} \Rightarrow b^2 = xc = x(x + y)$$

And,

$$\frac{a}{c} = \frac{y}{a} \Rightarrow a^2 = yc = y(x + y)$$

Hence,

$$a^2 + b^2 = x(x + y) + y(x + y) = (x + y)^2 = c^2$$

QED.

The way in which the assumption of Euclidean geometry sneaks into this proof is that the similarity of the triangles assumes that the angles of triangles always add to 180°, which is not true in general in non-Euclidean geometries.

The origin of Pythagoras's theorem in triangles belies its more general nature, namely as the distance between two points (D) in terms of the difference of their Cartesian coordinates, assuming Euclidean geometry. Thus, in any number n of dimensions we have,

$$D^2 = x_1^2 + x_2^2 + \cdots x_n^2$$

where x_j is the difference of the jth coordinates. This is known as the Euclidean metric and its assumption defines a Euclidean space.

Appendix B: Proof That e is Irrational

Euler (1737, but published seven years later) is credited with the first proof that e is irrational, by showing it has a representation as a non-terminating continued fraction. The rather lovely proof given here is due to Fourier. It uses the definition of e as an infinite series,

$$e = \sum_{r=0}^{\infty} \frac{1}{r!}$$

Proceeding by *Reductio ad Absurdum* we assume that there exist integers a and b such that $e = a/b$. Now consider,

$$x = b!\left(e - \sum_{r=0}^{b} \frac{1}{r!}\right) = b!\left(\frac{a}{b} - \sum_{r=0}^{b} \frac{1}{r!}\right) = a(b-1)! - \sum_{r=0}^{b} \frac{b!}{r!}$$

Note that each term in the last sum, b!/r!, is an integer because the sum extends only over $0 \le r \le b$. Consequently, x must be an integer.

But using the series for e we have,

$$x = b!\left(\sum_{r=0}^{\infty} \frac{1}{r!} - \sum_{r=0}^{b} \frac{1}{r!}\right) = \sum_{r=b+1}^{\infty} \frac{b!}{r!}$$

So,

$$x = \frac{1}{b+1} + \frac{1}{(b+1)(b+2)} + \frac{1}{(b+1)(b+2)(b+3)} + \cdots$$

And so,

$$x < \frac{1}{b+1} + \frac{1}{(b+1)^2} + \frac{1}{(b+1)^3} + \cdots$$

Using the familiar formula for the sum of a GP,

$$a + ar + ar^2 + \cdots = a/(1-r)$$

We thus find that,

$$x < \frac{1}{b+1}\left(\frac{1}{1 - \frac{1}{b+1}}\right) = \frac{1}{b}$$

Which means that $x < 1$ which contradicts the previous conclusion that x must be an integer. Thus the initial assumption, that there exists integers a and b such that $e = a/b$, must be false – and we conclude that e is irrational. QED.

References

Euler, Leonhard. (1744). *De fractionibus continuis dissertatio* [A dissertation on continued fractions] (PDF). Commentarii Academiae Scientiarum Petropolitanae. 9: 98–137.

de Stainville, Janot. (1815). *Mélanges d'Analyse Algébrique et de Géométrie* [A mixture of Algebraic Analysis and Geometry]. Veuve Courcier. pp. 340–341. (Fourier's proof)

Appendix C: Proof That π is Irrational

That π is irrational was first proved by Lambert in 1761 using continued fractions. The proof given below is of uncertain parentage, though it has become associated with Mary Cartwright (see Jeffreys, 1973), but one feels that 1945 is rather too late to be its true origin.

Consider the numbers I_n defined by the integrals,

$$I_n = \int_{-1}^{+1} (1 - x^2)^n \cos(\alpha x)\, dx$$

Integrating by parts, twice, assuming $n \geq 2$ the reader should confirm the following recurrence relation results,

$$\alpha^2 I_n = 2n(2n - 1)I_{n-1} - 4n(n - 1)I_{n-2}$$

For $n = 0$ and $n = 1$ we evaluate explicitly,

$$\alpha I_0 = 2 \sin \alpha \quad and \quad \alpha^3 I_1 = 4(\sin \alpha - \cos \alpha)$$

Then using the recurrence relation we get,

$$\alpha^5 I_2 = 2[(-8\alpha^2 + 24) \sin \alpha - 24 \cos \alpha]$$
$$\alpha^7 I_3 = 6[(-96\alpha^2 + 240) \sin \alpha + (96\alpha^2 - 240) \cos \alpha]$$

From which we see that the general form is,

$$\alpha^{2n+1} I_n = n! \, [P(\alpha) \sin \alpha + Q(\alpha) \cos \alpha]$$

where $P(\alpha)$ and $Q(\alpha)$ are polynomials in α of order less than $2n + 1$ and with integer coefficients. One can see from the recurrence relation how the factor of $n!$ arises, and this is essential to the proof. Putting $\alpha = \pi/2$ in this we have,

$$\left(\frac{\pi}{2}\right)^{2n+1} I_n \left(\frac{\pi}{2}\right) = n! \, P \left(\frac{\pi}{2}\right)$$

Now adopt *Reductio ad Absurdum* and assume that there are finite integers a and b such that $\pi = a/b$. Because P is a polynomial of order less than $2n + 1$ with integer coefficients it follows that

$$(2b)^{2n+1} P \left(\frac{\pi}{2}\right) = \frac{a^{2n+1}}{n!} I_n \left(\frac{\pi}{2}\right)$$

must be a non-zero integer, for all $n \geq 2$. But by inspection of the defining integral we see that,

$$I_n\left(\frac{\pi}{2}\right) < I_0\left(\frac{\pi}{2}\right) = \frac{4}{\pi}$$

because $I_n\left(\frac{\pi}{2}\right) = \int_{-1}^{+1}(1 - x^2)^n \cos\left(\frac{\pi}{2}x\right) dx$ and the factor of $(1 - x^2)^n$ in the integrand is less than, or equal to, 1 over the range of the integral, and the cos term is positive over the range of the integral. So we have been driven to the conclusion that the quantity,

$$\frac{a^{2n+1}}{n!} I_n\left(\frac{\pi}{2}\right)$$

where $I_n\left(\frac{\pi}{2}\right)$ is bounded above by a finite number (namely $4/\pi$), must be a non-zero integer for all n. But, however large a is assumed to be, this must tend to zero for large n, because.

$$\lim_{n\to\infty} \frac{a^{2n+1}}{n!} \to 0$$

So the expression which is claimed to be a non-zero integer cannot be as it tends to zero, thus establishing a contradiction from which we conclude there are no finite integers a and b such that $\pi = a/b$, i.e., that π is irrational. QED.

References

Lambert, Johann Heinrich. (1768). *Mémoire sur quelques propriétés remarquables des quantités transcendantes circulaires et logarithmiques*, see Pi: A Source Book (3rd Edition, 2014) by J.L.Berggren, J.Borwein & P.Borwein.

Jeffreys, Harold. (2011). *Scientific Inference* (Cambridge University Press 3rd edition)

Appendix D: Proof that e is Transcendental

Charles Hermite was the first to prove, in 1873, that e is transcendental. His proof underwent successive waves of simplification by a cast of heavyweights including Weierstrass, Hilbert, Hurwitz and Gordan. Here's the simplest version I know of. You will see we have to work a lot harder to prove transcendence than mere irrationality.

The proof is once again by *Reductio ad Absurdum*. Assume, contrary to what we suspect, that there exists a finite polynomial with integer coefficients of which e is a solution, thus,

$$a_m e^m + a_{m-1} e^{m-1} \dots a_1 e + a_0 = 0$$

where we can assume that a_0 is non-zero because, if it were zero, we could divide through by e to get a polynomial equation one order smaller with non-zero lowest coefficient – or repeat the process until we did.

Once again demonstrating that mathematics is an art – i.e., that one requires inspiration and creativity to produce proofs such as these – consider the function defined by,

$$f(x) = \frac{1}{(p-1)!} x^{p-1} (x-1)^p (x-2)^p \dots (x-m)^p$$

where p is some prime number, constraints on whose magnitude will emerge later. Note that $f(x)$ is defined with reference to the order of the polynomial equation obeyed by e, that is m. Note also that $f(x)$ is a polynomial of order $mp + p - 1$. Adopting the following notation for higher derivatives,

$$f^{(r)} \equiv \frac{\partial^r f}{\partial x^r} \text{ and } f^{(0)} = f$$

We define another polynomial of the same order by,

$$F(x) = \sum_{r=0}^{mp+p-1} f^{(r)}(x)$$

So, if we differentiate either f or F $mp + p$ times we get zero. We have,

$$\frac{d}{dx}[e^{-x} F] = e^{-x}[F^{(1)} - F] = -e^{-x} f$$

Hence it follows that, for any integer $j \in [0, m]$,

$$a_j \int_0^j e^{-x} f(x)dx = a_j[-e^{-x}F(x)]\big|_0^j = a_jF(0) - a_je^{-j}F(j)$$

Multiplying by e^j and summing over $j \in [0, m]$ gives,

$$\sum_{j=0}^m a_j e^j \int_0^j e^{-x} f(x)dx = F(0)\sum_{j=0}^m a_j e^j - \sum_{j=0}^m a_j F(j)$$

The first sum on the RHS is zero, by assumption. Substituting the definition of F in the second sum gives,

$$\sum_{j=0}^m a_j e^j \int_0^j e^{-x} f(x)dx = -\sum_{j=0}^m a_j \sum_{r=0}^{mp+p-1} f^{(r)}(j) \equiv Z$$

We are not too far off establishing a contradiction. The key is to recognise that, if $j \in [1, m]$, then one of the terms in the definition of $f(x)$ is $(x - j)^p$. Hence, the rth derivative, $f^{(r)}(x)$, will consist of terms each of which include a factor which is some power of $(x - j)$ providing that $r < p$. All those terms therefore vanish when evaluating at $x = j$, i.e., $f^{(r)}(j)$. On the other hand, for terms in the sum with $r \geq p$ the only non-zero terms will come from differentiating the $(x - j)^p$ factor exactly p times (the rest of the differentiation, if any, acting on the other factors). Hence all non-zero terms with $j \geq 1$ include a factor $p!$. Combining with the denominator of $(p - 1)!$ in the definition of $f(x)$ then leaves a factor of $\frac{p!}{(p-1)!} = p$.

Recalling that the coefficients a_j are integers, we conclude that the terms in Z with $j \geq 1$ are some integer divisible by p, i.e., of the form kp where k is an integer.

That leaves the $j = 0$ term in Z. The definition of $f(x)$ shows that $f^{(r)}(0)$ will be non-zero only if $r = p - 1$. The derivative of the x^{p-1} factor exactly $r = p - 1$ times brings down a factor of $(p - 1)!$ which cancels with the denominator in $f(x)$. Consequently the $j = 0$ term in Z equals $a_0(-1)^p(-2)^p \dots (-m)^p$. Hence, we conclude that,

$$Z = kp + a_0(-1)^p(-2)^p \dots (-m)^p \tag{D.1}$$

Now we take the prime p to be larger than both a_0 and m, from which it follows that Z is not divisible by p because the second term is not divisible by p but the first term is. (This is why p must be specified to be a prime, as well as larger than both a_0 and m). In particular, we conclude that Z cannot be zero.

Finally we now evaluate a bound on the integral which Z was derived to equal. Over the range $0 \le x \le m$, and hence over the range of the integral, it is clear from the definition that $|f(x)| \le m^{mp+p-1}/(p-1)!$. Hence,

$$\left| \sum_{j=0}^{m} a_j e^j \int_0^j e^{-x} f(x) dx \right| \le \sum_{j=0}^{m} |a_j e^j| \int_0^j \frac{m^{mp+p-1}}{(p-1)!} dx$$

because ignoring the sign of the integrand and integrating its magnitude can only make the integral larger, and ignoring the factor of e^{-x} in the integral similarly can only make the result larger as this factor is ≤ 1. The remaining integral is trivial and just gives j. Hence we conclude that,

$$|Z| = \left| \sum_{j=0}^{m} a_j e^j \int_0^j e^{-x} f(x) dx \right| \le \frac{m^{mp+p-1}}{(p-1)!} \left(\sum_{j=0}^{m} j |a_j e^j| \right)$$

The factor in round brackets is just some finite number. However, the factor in front of it tends to zero as we allow the prime $p \to \infty$. But this contradicts the earlier conclusion that Z could not be zero (or, to be more precise, it will inevitably be less than (D.1) for sufficiently large p). This establishes a contradiction and hence proves that the assumption that e is the solution to a finite polynomial equation with integer coefficients is false, i.e., e is transcendental. <u>QED</u>.

Appendix E: Proof that π is Transcendental

In 1882, Lindemann established that π was transcendental, using Hermite's proof for e as the inspiration. As far as I am aware, no one has come up with a proof which is more accessible to the non-specialist, and Lindemann's is far from easy. I include it here more to demonstrate how difficult these things can be, rather than because I expect many readers to follow the details. If you are interested in the details you might find it more congenial to start with the Mathologer's video on the subject (and even he admits it's one of his most challenging presentations).

I note in passing, though will not use it, that in 1885 Weierstrass obtained a generalisation of Lindemann's key result. A cut-down version of the full Lindemann-Weierstrass theorem states that if α is a nonzero complex number and e^{α} is algebraic, then α must be transcendental. It follows immediately that π is transcendental because we know that $e^{i\pi} = -1$. The modern approach to the full Lindemann-Weierstrass theorem is via *p-adic* numbers, see, for example, Matt Baker's blog.

(Note that the Lindemann-Weierstrass theorem does not imply that e^{π} is algebraic, nor that it is not!. But the transcendence of e^{π} follows from the theorem that if a is algebraic and b is irrational (and either might be complex) then a^b is transcendental. Because $e^{\pi} = (e^{i\pi})^{-i} = (-1)^{-i}$ it follows that e^{π} is transcendental, noting that $-i$ is irrational).

The proof is not quite as fiendish as it's made out, though it is certainly fiendishly cunning and intricate. Not the sort of thing that an ordinary pleb like me would have come up with in ten lifetimes. It relies on the properties of symmetric polynomials, so I start with these.

E.1 Symmetric Polynomials

A symmetric polynomial in two or more variables is a polynomial such that swapping any pair of variables leave it unchanged. Hence, for example,

$$x_1^2 x_2^3 + x_1^3 x_2^2$$

is a symmetric polynomial in two variables. (I am confining attention here to the field of complex numbers, though the concept of symmetric polynomials extends to other fields). In contrast,

$$x_1^2 x_2^3 + x_2^2 x_3^3 + x_1^2 x_3^3$$

is not a symmetric polynomial, but the following is a symmetric polynomial (in three variables),

$$x_1^2 x_2^3 + x_2^2 x_3^3 + x_1^2 x_3^3 + x_2^2 x_1^3 + x_3^2 x_2^3 + x_3^2 x_1^3$$

A particular set of symmetric polynomials are called the "elementary symmetric polynomials". For n variables there are n elementary symmetric polynomials. The general case is obvious from the following for $n = 4$,

$$e_1(x_1, x_2, x_3, x_4) = x_1 + x_2 + x_3 + x_4$$

$$e_2(x_1, x_2, x_3, x_4) = x_1 x_2 + x_1 x_3 + x_1 x_4 + x_2 x_3 + x_2 x_4 + + x_3 x_4$$

$$e_4(x_1, x_2, x_3, x_4) = x_1 x_2 x_3 + x_1 x_2 x_4 + x_1 x_3 x_4 + x_2 x_3 x_4$$

$$e_4(x_1, x_2, x_3, x_4) = x_1 x_2 x_3 x_4$$

This establishes a link with the coefficients of a single-variable polynomial when the above $x_1, x_2 \ldots x_n$ are now interpreted as its roots. For this purpose we assume a single-variable polynomial in monic form, i.e., with leading coefficient unity,

$$t^n + a_{n-1} t^{n-1} + \cdots a_1 t + a_o$$

This can be written in terms of its roots as,

$$(t - x_1)(t - x_2) \ldots (t - x_n)$$

Expanding shows that the coefficients a_r are just the elementary symmetrical polynomials in the roots, $x_1, x_2 \ldots x_n$, except with alternating signs, i.e.,

$$a_{n-1} = -e_1(x_1, x_2, \ldots, x_n)$$

$$a_{n-2} = e_2(x_1, x_2, \ldots, x_n)$$

$$a_{n-3} = -e_3(x_1, x_2, \ldots, x_n)$$

$$a_{n-4} = e_4(x_1, x_2, \ldots, x_n)$$

etc., to,

$$a_1 = (-1)^{n-1} e_{n-1}(x_1, x_2, \ldots, x_n)$$

$$a_0 = (-1)^n e_n(x_1, x_2, \ldots, x_n)$$

In particular, therefore, if the coefficients a_r are all rational, then, whilst we can only conclude that the roots $x_1, x_2 \ldots x_n$ are all algebraic, the combinations of them given by the elementary symmetric polynomials, $e_r(x_1, x_2, \ldots, x_n)$, are rational.

This observation gains greater power due to the theorem that any symmetric polynomial in n variables with rational coefficients can be expressed as a polynomial in the elementary symmetric polynomials, $e_r(x_1, x_2, \ldots, x_n)$. (I will not provide the proof here but you can look it up in any standard source). Here's an example for two variables,,

$$x_1^3 + x_2^3 + 8 = e_1(x_1, x_2)^3 - 3e_1(x_1, x_2)e_2(x_1, x_2) + 8$$

That is,

$$x_1^3 + x_2^3 + 8 = (x_1 + x_2)^3 - 3(x_1 + x_2)x_1 x_2 + 8$$

Hence, any symmetric polynomial in the n roots of a single-variable monic polynomial, f, can be expressed as a polynomial in the coefficients of f.

The converse also holds: if a polynomial, θ, in the n roots of a single-variable monic polynomial, f, can be expressed as a polynomial in the coefficients of f then θ is a symmetrical polynomial in the roots of f.

E.2 The Proof

This follows Ian Stewart's presentation of Lindemann's proof (but I've corrected what I think was an error, so you will need to check my proof carefully). We proceed as usual by *Reductio ad Absurdum*. Hence, assume there is a polynomial with rational coefficients one of whose roots is π. It follows there is also a polynomial, θ_1, with rational coefficients one of whose roots is $i\pi$. Let the roots of θ_1 be called $\alpha_1, \alpha_2, \alpha_3 \ldots \alpha_n$ where $\alpha_1 = i\pi$. Hence we have,

$$(e^{i\alpha_1} + 1)(e^{i\alpha_2} + 1)(e^{i\alpha_3} + 1) \ldots (e^{i\alpha_n} + 1) = 0 \tag{E.1}$$

Expanding the brackets in (E.1) yields terms whose exponents are of the form $\alpha_i + \alpha_j$ and $\alpha_i + \alpha_j + \alpha_k$ and $\alpha_i + \alpha_j + \alpha_k + \alpha_l$, etc., for all combinations of $i, j, k, l, \ldots$. We seek to demonstrate the existence of a polynomial, $\tilde{\theta}$, with integer coefficients whose roots are these combinations of the roots of θ_1, including the n roots α_i themselves, so that θ will be required to have $2^n - 1$ roots.

Consider the elementary symmetric polynomials formed from the $n(n-1)/2$ summed pairs $\alpha_i + \alpha_j$. These will also be symmetric in α_i themselves, and hence can be expressed as a polynomial in the coefficients of θ_1, which are rational. We have thus shown that all the elementary symmetric polynomials formed from the set of all summed pairs $\alpha_i + \alpha_j$ are rational. But these elementary symmetric polynomials are (modulo their sign) just the coefficients of a polynomial, θ_2, whose $n(n-1)/2$ roots are the summed pairs $\alpha_i + \alpha_j$ and so the polynomial, θ_2, has rational coefficients.

In the same way a polynomial with rational coefficients, θ_3, can be constructed whose $n(n-1)(n-2)/6$ roots are all possible summed triples $\alpha_i + \alpha_j + \alpha_k$, and so on. Hence,

$$\tilde{\theta} = \theta_1 \theta_2 \theta_3 \dots \theta_n \qquad (E.2)$$

is a polynomial with rational coefficients whose $2^n - 1$ roots are the exponents in (E.1) when expanded. Multiplying (E.2) by a suitable integer makes it a polynomial with integer coefficients. Some of its roots might be zero, but not all can be because $\alpha_1 = i\pi$ by assumption, and hence we can divide (E.2) by a suitable integral power of its variable to create a polynomial θ whose roots are the non-zero exponents in (E.1) when expanded. Hence (E.1) can now be written,

$$e^{\beta_1} + e^{\beta_2} + \cdots + e^{\beta_r} + k = 0 \qquad (E.3)$$

where β_1 to β_r are the non-zero exponents in (E.1) when expanded and k is a positive integer of at least 1 (accounting for the potential for $k-1$ exponents to be zero, plus one because of the term 1 in the expansion of (E.1)). The polynomial θ, suitably modified from $\tilde{\theta}$, above, can be written,

$$\theta = cx^r + c_1 x^{r-1} + \cdots + c_r \qquad (E.4)$$

where the c_i are integers. Note that its order is the same as the number of exponentials in (E.3) because β_1 to β_r are the roots of (E.4), which are all non-zero, so that both c and c_r are non-zero. In a similar way to how we proceeded in proving the transcendent nature of e in Appendix D, we define,

$$f(x) = \frac{c^s x^{p-1} [\theta(x)]^p}{(p-1)!} \qquad (E.5)$$

where $s = rp - 1$ and p is some prime number (which will later be chosen to be sufficiently large). Note this differs from Ian Stewart. Also define,

$$F(x) = f(x) + f^{(1)}(x) + f^{(2)}(x) + \cdots f^{(s+p)}(x)$$

where the superscript denotes the corresponding derivative. Note that (E.5) shows $f(x)$ to be a polynomial of order $pr + p - 1$ so that $f^{(s+p+1)}(x)$ is identically zero. As in Appendix D we have,

$$\frac{d}{dx}[e^{-x}F] = e^{-x}[F^{(1)} - F] = -e^{-x}f$$

Hence, integrating, changing the integration variable on the RHS to μx, and multiplying throughout by e^x, gives,

$$F(x) - e^x F(0) = -x \int_0^1 e^{(1-\mu)x} f(\mu x) d\mu$$

We now replace x by each of the β_1 to β_r in turn and sum over the result to give,

$$\sum_{i=1}^r F(\beta_i) + kF(0) = -\sum_{j=1}^r \beta_j \int_0^1 e^{(1-\mu)\beta_j} f(\mu\beta_j) d\mu \qquad \text{(E.6)}$$

Note that the LHS of (E.6) has used (E.3). The next step is to show that, for a sufficiently large prime p, the LHS of (E.6) is a non-zero integer.

By definition we have $\theta(\beta_i) = 0$ for all $1 \le i \le r$ and hence, from the definition (E.5), $f(\beta_i) = 0$ and all its derivatives below the pth derivative are also zero, $f^{(t)}(\beta_i) = 0$ for $1 \le t \le p - 1$.

The derivatives $f^{(t)}(\beta_i)$ for $t \ge p$ all have a factor of p because the $[\theta(x)]^p$ term in (E.5) must be differentiated p times to avoid being zero, and this brings down a factor of $p!$, then division by $(p-1)!$, as in (E.5), leaves a factor of p, and this factors a non-zero term. As $f^{(t)}(\beta_i)$ for a given i is a polynomial in the chosen β_i alone, it follows that summing over i from 1 to r, i.e.,

$$\sum_{i=1}^r f^{(t)}(\beta_i) \qquad \text{(E.7)}$$

is a symmetric polynomial in the β_1 to β_r. Furthermore, for $t \ge p$, as f is a polynomial of order $pr + p - 1$, the sum (E.7) is a non-zero symmetric polynomial of order $\le pr - 1 = s$. But the β_i are the roots of θ, (E.4), and hence (E.7) is also a polynomial of order $\le s$ in the monic coefficients c_i/c. The factor of c^s in the definition (E.5) then turns all the powers of c_i/c into integers. Finally, then, we have

$$\sum_{i=1}^r f^{(t)}(\beta_i) = k_t p \qquad \text{(E.8)}$$

for some integers k_t. Hence, in (E.6), $\sum_{i=1}^{r} F(\beta_i) = k_F p$ where k_F is also an integer. Now turn attention to $F(0)$. If we differentiate (E.5) less than $p - 1$ times, then the result is clearly zero at $x = 0$. Differentiating exactly $p - 1$ times gives $f^{(p-1)}(0) = c^s c_r^p$, because $\theta(0) = c_r$, and hence $f^{(p-1)}(0)$ is a non-zero integer. Finally, differentiating (E.5) p or more times will involve a factor of p times derivatives of $\theta(x)$ evaluated at $x = 0$, which are therefore integers, times $c^s c_r^p$, and so the overall result is some integer times p.

Putting all this together, the LHS of (E.6) is of the form,

$$\sum_{i=1}^{r} F(\beta_i) + kF(0) = Kp + kc^s c_r^p \qquad \text{(E.9)}$$

for some integer K. Now k, c, c_r are all non-zero. If we take p to be a large enough prime, namely $p > \max(k, c, c_r)$, then (E.9) is an integer not divisible by p and hence is also non-zero. All the effort so far has been to derive this result. We now seek to show that this contradicts the RHS of (E.6) by showing that the RHS of (E.6) can be arbitrarily small, tending to zero as $p \to \infty$. This last part closely follows the proof of the transcendence of e.

From (E.5) we have, for the integration range $0 \leq \mu \leq 1$,

$$f(\mu\beta_j) = \frac{|c|^s |\beta_j|^{p-1} m(j)^p}{(p-1)!}$$

where $m(j)$ is the maximum value of $|\theta(\mu\beta_j)|$ over the integration range. Hence the RHS of (E.6) obeys,

$$\left| -\sum_{j=1}^{r} \beta_j \int_0^1 e^{(1-\mu)\beta_j} f(\mu\beta_j) d\mu \right| \leq \sum_{j=1}^{r} \frac{|c|^s |\beta_j|^p |m(j)|^p}{(p-1)!} B$$

(E.10)

where,

$$B = \left| max_j \left(\int_0^1 e^{(1-\mu)\beta_j} d\mu \right) \right|$$

But however large the fixed quantities $|\beta_j|$ and $|m(j)|$ might be, the ratio $\frac{|\beta_j|^p |m(j)|^p}{(p-1)!}$ will become arbitrarily small, tending to zero, as $p \to \infty$, and hence so does the (modulus of the) RHS of (E.6). But the LHS of (E.6) has been shown to be a non-zero integer, which established the desired contradiction and hence proves that π is transcendental. **QED**.

E.3 References

Baker, M. (2015). *A p-adic proof that π is transcendental.*
https://mattbaker.blog/2015/03/20/a-p-adic-proof-that-pi-is-transcendental/

Mathologer (2018). *The Proof: e and π are transcendental.*
https://www.youtube.com/watch?v=WyoH_vgiqXM

Stewart, I. (1973). *Galois Theory.* Chapman and Hall. The link is to the first edition. You may prefer the most recent edition.

Appendix F: Algebraic Fields and the Unconstructible Trisector

I acknowledge David Bailey as a source for much of this Appendix. You can consult that link for more details, e.g., detailed proofs. This Appendix can be used as a preliminary orientation exercise before tackling Galois Theory in its full generality, for which see Stewart (1973). Firstly some definitions.

F.1 Fields

A field, F, is defined as a set which is closed under two binary operations (sum and product) and such that both sum and product have a "unit element" (call them 0 and 1 respectively) which leave all members of F invariant, i.e., $x + 0 = x$ and $x \times 1 = x$ for all $x \in F$. All elements of F have an inverse under both sum and product, except the unit under summation, i.e., $-x$ exists in F such that $x + (-x) = 0$ and $1/x$ exists in F such that $x \times (1/x) = 1$ except for $x = 0$. Both summation and multiplication are required to commute, $x + y = y + x$ and $x \times y = y \times x$ for all x, y in F. Both addition and multiplication are required to be associative, $(x + y) + z = x + (y + z)$ and $(x \times y) \times z = x \times (y \times z)$. Finally, multiplication is distributive over addition, $x \times (y + z) = x \times y + x \times z$.

In short, a field is defined so as to share the basic algebraic properties of numbers (up to and including complex numbers, but not quaternions or octonions).

For example, the rational numbers Q are a sub-field of the real number field R, written $Q \subset R$. This can equivalently be expressed as "R is an extension of Q" and written as R/Q, read as "R is a field extension over Q".

Note that the integers are not a field (they have no multiplicative inverse, but they do form a slightly weaker structure, a commutative ring). However the finite set of integers $[0, p - 1]$ forms a field under arithmetic modulo p.

F.2 Irreducible Polynomials

A polynomial with integer coefficients is said to be irreducible if it cannot be factored into two or more polynomials of lesser degree, also with integer coefficients. Hence $x^2 - 3x - 2$ is irreducible whereas $x^2 - 3x + 2$ is reducible as it equals $(x - 2)(x - 1)$. Similarly, $x^3 - 2$ is irreducible.

F.3 The Degree of an Algebraic Number / Minimal Polynomial

If a number is algebraic there exists a polynomial with integer coefficients of which it is a root. The smallest order of polynomial for which it is a root is the degree of the algebraic number. If its degree is n then the corresponding polynomial, $a_n x^n + a_{n-1} x^{n-1} + \cdots + a_0$ where the a_k are integers is said to be the minimal polynomial of the number. [Note that neither a_0 nor a_n can be zero because, if so, the polynomial would not be minimal]. In other words, the order of the irreducible polynomial with x as a root is the degree of x.

F.4 The Degree of a Field Extension

Suppose that some field, S, is an extension of the rational field, $Q \subset S$, or equivalently S/Q. Further suppose that every element of S can be written as a linear sum of terms in a certain set of elements within S, namely $\{s_i, i \in [1, m]\}$, with rational coefficients. Hence, every element in S can be written as $q_1 s_1 + q_2 s_2 + \cdots q_m s_m$. Essentially, S is a vector space over the rationals with $\{s_i, i \in [1, m]\}$ as basis vectors. To complete the definition, and the analogy with vector spaces, the basis vectors must be linearly independent with respect to Q. That is, if $q_1 s_1 + q_2 s_2 + \cdots q_m s_m = 0$ it must be that all the q_i are zero.

Note that, as with other vector spaces, if m were chosen too small then not every element of S could be expressed as $q_1 s_1 + q_2 s_2 + \cdots q_m s_m$. On the other hand, if m were chosen too large then not all the $\{s_i, i \in [1, m]\}$ could be linearly independent. Confining attention to the finite case, there is one m which meets both requirements. This is the degree of the field extension S over Q.

An example is generated by the basis $\left\{1, 2^{\frac{1}{3}}, 2^{\frac{2}{3}}\right\}$, the field being over the rationals. This basis is linearly independent wrt Q because there are no rational numbers q_2, q_3 such that $2^{\frac{1}{3}} q_2 + 2^{\frac{2}{3}} q_3$ is rational. This may be proved simply using the machinery we already have in place. $2^{\frac{1}{3}}$ is clearly a root of $x^3 - 2$, which is an irreducible polynomial, hence the degree of the algebraic number $2^{\frac{1}{3}}$ is 3. That means there can be no quadratic of which $2^{\frac{1}{3}}$ is a root. Hence, if an element of the field, i.e., $q_1 + q_2 2^{\frac{1}{3}} + q_3 2^{\frac{2}{3}}$, is zero then, as this is quadratic, we must have $q_1 = q_2 = q_3 = 0$.

F.5 Fields Generated by an Algebraic Number

A generalisation of the above example is a field over the rationals defined by some chosen algebraic number, α, of degree $m \geq 2$, and defined as having the general element $q_0 + q_1\alpha + q_2\alpha^2 + \cdots q_{m-1}\alpha^{m-1}$. Such a set of elements is clearly closed under addition, but it is not so obvious that it is closed under multiplication. This because the product of two such elements would seem to involve all powers of α up to and including α^{2m-2}, so how can this be reduced to a form only including powers up to α^{m-1}? The answer lies in the fact that we are given that the degree of α is m so there are rational numbers (or integers if you prefer) such that $r_0 + r_1\alpha + r_2\alpha^2 + \cdots r_m\alpha^m$ is zero. Hence there are rational numbers q_i such that α^m can be written as a field element $\alpha^m = q_0 + q_1\alpha + q_2\alpha^2 + \cdots q_{m-1}\alpha^{m-1}$. Multiplying that by α gives an expression for α^{m+1} as $q_0\alpha + q_1\alpha^2 + q_2\alpha^3 + \cdots q_{m-1}\alpha^m$ and the last term can be replaced by the above expression for α^m to give an expression for α^{m+1} as a polynomial of order (at most) $m - 1$. Repeating this process finds all the $\alpha^n, n \in [m, 2m - 2]$ as polynomials of order $m - 1$ at most.

Hence, the candidate field is indeed closed under multiplication. The other field properties can also be confirmed. So the set of elements, S, defined by $q_0 + q_1\alpha + q_2\alpha^2 + \cdots q_{m-1}\alpha^{m-1}$ and generated by any algebraic number, α, of degree $m \geq 2$, is indeed a field. Such a field is an extension over the field of rationals, $Q \subset S$, or equivalently S/Q.

Moreover, the degree of a field extension of the rationals defined in this way is the number of terms in $q_0 + q_1\alpha + q_2\alpha^2 + \cdots q_{m-1}\alpha^{m-1}$, the most general element. Hence, the degree of a field extension generated by an algebraic number of degree m is m. In other words, the notions of algebraic degree and field extension degree coincide when the field is generated by an algebraic number.

F.6 The Multiplicative Degrees Theorem

Suppose there are three fields $Q \subset S \subset T$, where the degree of S/Q is a and the degree of T/S is b (both finite), then the degree of T/Q is ab. For a proof see

https://proofwiki.org/wiki/Degree_of_Field_Extensions_is_Multiplicative

F.7 The Chosen Angle

To prove that an arbitrary angle cannot be trisected with a rule and a pair of compasses it suffices to show that any particular angle cannot be trisected. We choose the angle $\pi/3$ so that the trisected angle would be $\pi/9$. Using the standard trig formula $\cos 3\theta = 4\cos^3\theta - 3\cos\theta$ and putting $\beta = 2\cos\frac{\pi}{9}$ gives $\beta^3 - 3\beta - 1 = 0$.

In a sense to be defined below, constructing the trisector of $\pi/3$ is equivalent to constructing a length equal to β from an initial arbitrary unit length. But it is easily seen that the cubic polynomial $\beta^3 - 3\beta - 1$ is irreducible (proof follows). So we see that the task is to construct a number which is algebraic of degree 3. It will be seen that this is impossible, and hence that constructing the trisector of an arbitrary angle is impossible.

F.8 The Irreducibility of $\beta{\char`\^}3 - 3\beta - 1$

One way of showing this is to use the Rational Root Theorem, which says that if $x = p/q$ is a rational root of $a_n x^n + a_{n-1}x^{n-1} + \cdots + a_0$ where the a_k are integers, then p divides a_0 and q divides a_n. The proof is elementary and left as an exercise for the reader.

Now, if $\beta^3 - 3\beta - 1$ can be factored into terms with integer coefficients, then it must have at least one rational root because one of the factors must be linear (i.e., of degree 1) with integer coefficients. But then p would divide 1, and so would q, so both must be 1. But that would require $\beta = 1$ to be a root, which it clearly is not. Hence, by *Reductio ad Absurdum*, $\beta^3 - 3\beta - 1$ cannot be factored into terms with integer coefficients, i.e., it is irreducible.

F.9 Constructable Lengths

We are considering the Euclidean plane. "Construction" refers to the use of a pair of compasses plus a rule (which means a straight edge, no measuring distances!). Firstly, note that if we can construct the trisector of an angle, 3θ, then we can also construct two points which are a distance $\cos\theta$ apart (simply by dropping a perpendicular, which is constructable). Hence, if the latter is impossible, so is the former. So, we seek to prove that construction of a length β is impossible (starting from some arbitrary length which functions as a unit length).

What points can be constructed? They are all defined by intersections, of which there are three kinds: a line intersecting another line, a line intersecting a circle, or two mutually intersecting circles. We want to know what order of polynomial defines the coordinates of the new point.

For an intersection of two lines it is clear that the coordinates of the new point will be specified by a linear equation, i.e., the "polynomial" in question is of order 1.

For both a line intersecting a circle and two intersecting circles, the new point is defined by a quadratic. For a line $y = mx + c$ and a circle $(x - a)^2 + (y - b)^2 = r^2$ the quadratic defining x is,

$$(x - a)^2 + (mx + c - b)^2 = r^2$$

with a similar expression for y. For the intersection of the circle with a second circle I leave it as an exercise to confirm that the intersection points are again given by quadratics. (This is physically obvious as there are at most two intersection points).

We are starting from the points $(0,0)$ and $(0,1)$, from which we can easily construct the angle $\pi/3$ from the point which forms an equilateral, unit-sided, triangle with the starting points, specifically the point $\left(\frac{1}{2}, \frac{\sqrt{3}}{2}\right)$.

Construction can obviously add or subtract lengths. But it can also multiply or divide lengths. This is not so obvious, but here is a scheme for doing so. Referring to Figure 5.1, mark the two lengths to be multiplied on the same line, from the same origin, x and y. Draw an arbitrary sloping line from the origin. Construct the perpendicular at distance 1 from the origin. Draw the line from where this intersects the sloping line to y. Now construct a line parallel to this through the point where the perpendicular from x intersects the sloping line. This line intersects the original line at a distance xy as is readily shown. Division can be done by starting with x and xy and constructing y. Hence we have shown that the set of constructable points is closed under addition and multiplication, and these operations have constructable inverses.

Figure 5.1: *Taken from* <u>geometric construction - Representing the multiplication of two numbers on the real line - Mathematics Stack Exchange</u>

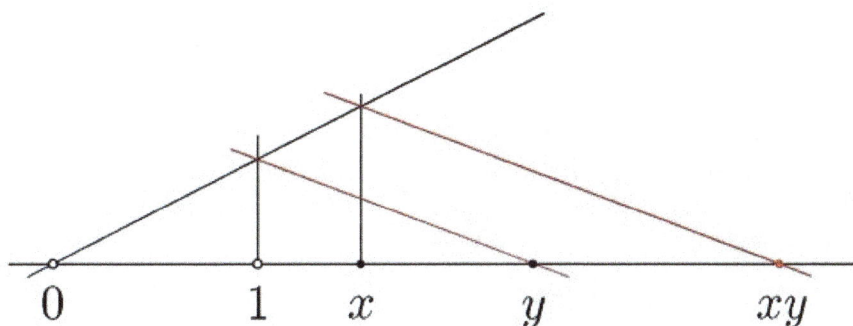

It is becoming clear that the set of points which are constructable forms a field over the set of points already constructed. But we have not yet quite characterised the field. Note that square-roots of lengths are also constructible. A scheme for doing this is illustrated by Figure F.2, the cited source explains the procedure thus,

"If you have a segment AB, place the unit length segment on the line where AB lies, starting with A and in the direction opposite to B; let C be the other point of the segment. Now draw a semicircle with diameter BC and the perpendicular to A; this line crosses the semicircle in a point D. Now AD is the square root of AB."

Figure F.2. *Taken from* geometry - Compass-and-straightedge construction of the square root of a given line? - Mathematics Stack Exchange. $e = \sqrt{a}$

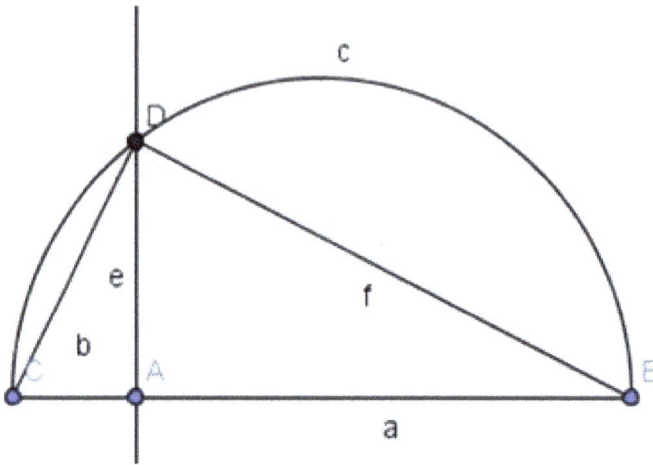

F.10 Finally – The Proof

We have already seen that all new points are defined by a polynomial of no higher order than quadratic. Suppose we denote by F_n the field defined by the set of points acquired (or acquirable) after n constructed intersections. The next generation of points, after $n + 1$ constructed intersections, will be defined by quadratics over F_n. That is, the field F_{n+1} is an extension of F_n, and, moreover, the degree of F_{n+1}/F_n is 2 (i.e., quadratic).

Before any constructions have been made, the starting points, (0,0) and (0,1), are defined by integers. Hence F_1/Q has degree 2. It immediately follows from the Multiplicative Degrees Theorem, above, that the degree of F_n/Q is 2^n.

But we seek to construct the length β, and we have seen that this is defined as a root of the irreducible cubic $\beta^3 - 3\beta - 1$, i.e., β is an algebraic number of degree 3. Again by the Multiplicative Degrees Theorem, any field extension containing β can only have degrees which are multiples of 3.

But we have seen that the only constructible numbers have degree 2^n for some integer n, and this cannot be a multiple of 3. Hence β is not constructible. Hence the trisector of angle $\pi/3$ is not constructible. Hence the trisector of an arbitrary angle is not constructible. **<u>QED</u>**.

F.11 References

Bailey, D.H. (2018). *Simple proofs: The impossibility of trisection.* September 29th, 2018. https://mathscholar.org/2018/09/simple-proofs-the-impossibility-of-trisection/

Stewart, I. (1973). *Galois Theory.* Chapman and Hall. The link is to the first edition. You may prefer the most recent edition.

Appendix G: The Solution of the Cubic

We seek the solutions of,

$$az^3 + bz^2 + cz + d = 0 \qquad \text{(G.1)}$$

Here I want to illustrate how the solution can be found, not merely to write it down. Invent a crazy number system, like complex numbers but with three parts instead of two, i.e., with two different 'imaginary' parts denoted by j and k. The multiplication table is,

	1	j	k
1)	1	j	k
j)	j	k	1
k)	k	1	j

$$\text{(G.2)}$$

Of course there is no such sensible number system really[1], as Hamilton found out. But we'll not let that stop us. Write,

$$z = x + jy + kv \qquad \text{(G.3)}$$

Substitute (G.3) into (G.1) and demand that the three parts of the result (the real part and the two different imaginary parts) are individually zero, just as they would have to be if (G.2) were a sensible number system[2]. This gives the three equations,

$$a\left(x^3 + y^3 + v^3 + 6xyv\right) + b\left(x^2 + 2yv\right) + cx + d = 0 \qquad \text{(G.4)}$$

$$3a\left(xv^2 + x^2y + y^2v\right) + b\left(v^2 + 2xy\right) + cy = 0 \qquad \text{(G.5)}$$

$$3a\left(x^2v + xy^2 + yv^2\right) + b\left(y^2 + 2xv\right) + cv = 0 \qquad \text{(G.6)}$$

Multiplying (G.5) by v and (G.6) by y and subtracting or adding the results gives,

---

[1] Finite dimensional division algebras over the reals must have dimension 1, 2, 4 or 8. The only possible associative division algebras of dimension 1, 2 and 4, up to isomorphism, are the reals, the complex numbers and the quaternions. I'm not sure that the same uniqueness can be claimed for dimension 8, though the octonions are the only alternative division algebra of dimension 8 and the only normed algebra over the reals of dimension 8.

[2] What we are really doing here, in disguise, is appealing to the field generated by the algebraic number $\omega = exp\{2i\pi/3\}$, as explained in §F.4,5, but we don't know that yet!

$$(3ax+b)\left(v^3 - y^3\right) = 0 \tag{G.7}$$

$$(3ax+b)\left(v^3 + y^3\right) + 6a\left(x^2 yv + y^2 v^2\right) + 4bxyv + 2cyv = 0 \tag{G.8}$$

(G.7) implies that either $v^3 = y^3$ or that,

$$x = -\frac{b}{3a} \tag{G.9}$$

We choose to follow the consequences of this latter possibility. For a start it simplifies (G.8) to,

$$3a\left(x^2 yv + y^2 v^2\right) + 2bxyv + cyv = 0 \tag{G.10}$$

which implies that $v = 0$ or $y = 0$ or,

$$3a\left(x^2 + yv\right) + 2bx + c = 0 \tag{G.11}$$

Again we choose to follow the consequences of the latter possibility. Substituting for x from (G.9) this gives,

$$yv = \frac{b^2 - 3ac}{\left(3a\right)^2} \equiv D \tag{G.12}$$

(G.9) and (G.12) have been derived from (G.5) and (G.6) only, so we now need to use (G.4). Substituting (G.9) into (G.4) it becomes,

$$a\left(y^3 + v^3\right) = -\left(ax^3 + bx^2 + cx + d\right) = -aE \tag{G.13}$$

where,
$$E = \frac{2b^3 - 9abc + 27a^2 d}{\left(3a\right)^3} \tag{G.14}$$

Combining (G.12) and (G.13) we have,

$$y^6 + Ey^3 + D^3 = 0 \tag{G.15}$$

This yields the solutions,

$$y^3 = Y_{\pm} = \left[-E \pm \sqrt{E^2 - 4D^3}\right]/2 \tag{G.16}$$

At first sight we seem to have a superfluity of solutions, since there are three cube roots of each of Y_+ and Y_-, apparently making six solutions in all. However, it is readily seen from (G.16) that $Y_+ Y_- = D^3$, and comparison with (G.12) shows that we can therefore interpret,

$$y = \left\{1, \omega, \omega^2\right\}\left(Y_+\right)^{\frac{1}{3}} \quad \text{and} \quad v = \left\{1, \omega^2, \omega\right\}\left(Y_-\right)^{\frac{1}{3}} \tag{G.17}$$

where ω is a complex cube-root of unity, $\omega = \exp\{2\pi i/3\} = (-1 + \sqrt{3}i)/2$. Each solution for y must be matched with the corresponding solution for v in order for (G.12) to be respected.

There remains the interpretation of (G.3), without which we have merely found the solutions in terms of our crazy number system. We now note that setting $j = \omega$ and $k = \omega^2$ respects the multiplication table, (G.2). Hence, finally we have the three solutions to the general cubic,

$$z = -\frac{b}{3a} + \{1, \omega, \omega^2\}(Y_+)^{1/3} + \{1, \omega^2, \omega\}(Y_-)^{1/3} \tag{G.18}$$

where $Y_\pm$ are given by (G.16) using (G.12) and (G.14) for D and E. **QED**.

Note that we could not simply set $j = \omega$ and $k = \omega^2$ from the start because we would not then have arrived at the three equations (G.4), (G.5), (G.6). Instead, working in ordinary complex numbers, we would have had just two equations. So our ruse of using the crazy number system is to get the extra equation. The fact that this number system ultimately does not make sense does not undermine the solution, (G.18), since it is now guaranteed that substitution of any of the three solutions, (G.18), into the LHS of (G.1) will reveal it to be identically zero.

Note that there is always one real root, as we know there has to be. If Y_+ and Y_- are real and different, then there is only one real root, which is the first in the order listed in (G.18). If Y_+ and Y_- are real and equal, or if they are complex (and hence conjugates) then there are three real roots.

The solution for both the cubic and the quartic polynomials was first published in 1545 by Cardano in his famous *Ars Magna*. (Cardano's student Ferrari found the solution to the quartic five years earlier, but in a form that required the solution to the cubic which had not yet been found at that time).

Appendix H: Proof of Green's Theorem

Green's Theorem says,

$$\oint_{\delta C}(gdx + hdy) = \iint_C \left(\frac{\partial h}{\partial x} - \frac{\partial g}{\partial y}\right)dxdy$$

providing the derivatives and integrals exist, and noting that the boundary (path) integral is conducted counter-clockwise. Consider the region C bounded by δC below,

Figure H.1

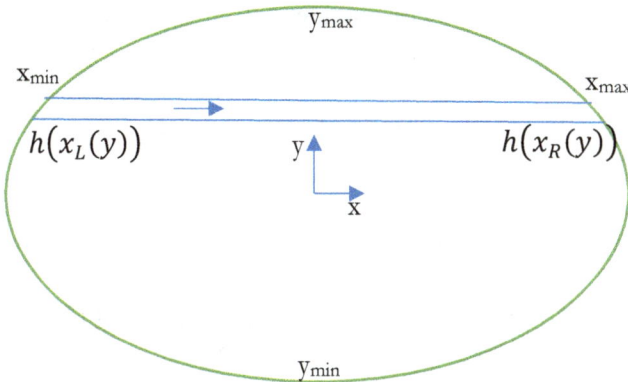

We seek first to establish the theorem when $g = 0$, i.e., that,

$$\oint_{\delta C} hdy = \iint_C \frac{\partial h}{\partial x}dxdy$$

As the diagram shows we may split the area integral on the RHS into narrow strips of constant y. The contribution to the area integral of this single strip can be found because the integral over x of $\frac{\partial h}{\partial x}dx$ is just Δh, which is $h(x_R(y)) - h(x_L(y))$, where those quantities are the value of the function h on the right-hand and left-hand boundaries at this y value. (Note that the blue arrow indicates the sign of this integral, being from x_{min} to x_{max}. The area integral now requires us to carry out the y-integral, from y_{min} to y_{max}, giving,

$$RHS = \int_{y_{min}}^{y_{max}} [h(x_R(y)) - h(x_L(y))]dy$$

But this can equally be written,

$$RHS = \int_{y_{min}}^{y_{max}} h(x_R(y))dy + \int_{y_{max}}^{y_{min}} h(x_L(y))dy$$

whereupon we note that this is simply the LHS of the equation to be proved, since the first term is the counter-clockwise path integral on the right-hand part of the boundary and the second term is the counter-clockwise path integral on the left-hand part of the boundary.

The proof of the other half of Green's Theorem, i.e.,

$$-\oint_{\delta C} g\,dx = \iint_C \frac{\partial g}{\partial y}\,dx\,dy$$

proceeds in identical fashion, considering vertical y-integration strips at constant x, then integrating over x, thus,

$$RHS = \int_{x_{min}}^{x_{max}} [g(y_T(x)) - g(y_B(x))]dx$$

where y_T, y_B are the top and bottom of the vertical strips defining the top and bottom parts of the boundary respectively. This can be rewritten as,

$$RHS = -\int_{x_{max}}^{x_{min}} g(y_T(x))dx - \int_{x_{min}}^{x_{max}} g(y_B(x))dx$$

showing how an overall minus sign occurs, and establishing that this equals the LHS of the equation to be proved. Note that the sign is dictated by the requirement that the path integral is conducted counter-clockwise. Were we to reverse the limits of integration in the above expression, and thus to remove the minus signs, the resulting path integral would be clockwise. QED.

Appendix I: Bayes' Theorem

You can spot a Bayesian – they have this printed on their tee-shirts,

$$P(A|B)P(B) = P(B|A)P(A)$$

where,

$P(A)$ is the probability of event A;

$P(B)$ is the probability of event B;

$P(A|B)$ is the conditional probability of event A assuming that event B has occurred (or "is true");

$P(B|A)$ is the conditional probability of event B assuming that event A has occurred (or "is true").

The truth of Bayes' Theorem is most easily seen from the associated Venn diagram, Figure H.1. Let $N(U)$ be the totality of events in the "universe" (i.e., the whole population). Similarly, $N(A)$ is the number of A-events, $N(B)$ is the number of B-events, and $N(A \cap B)$ is the number of events which are both of type A and B (i.e., within the intersection of the A and B sets). The frequentist definition the tells us that,

$$P(A) = \frac{N(A)}{N(U)} \quad \text{and} \quad P(B) = \frac{N(B)}{N(U)}$$

However, the conditional probabilities are defined wrt a denominator which is the number of conditional events, i.e.,

$$P(A|B) = \frac{N(A \cap B)}{N(B)} \quad \text{and} \quad P(B|A) = \frac{N(A \cap B)}{N(A)}$$

These imply that,

$$P(A|B)N(B) = N(A \cap B) = P(B|A)N(A)$$

Dividing by $N(U)$ this is equivalent to,

$$P(A|B)P(B) = P(B|A)P(A) \qquad \text{QED.}$$

There is nothing to object to about Bayes' Theorem. Controversies arise in applications, especially in the context of criminal justice or the chance of a patient having a serious disease. It's worth driving this home with some numerical examples because they provide further illustrations of the counter-intuitive nature of probabilities.

Fig H.1: Bayes' Theorem

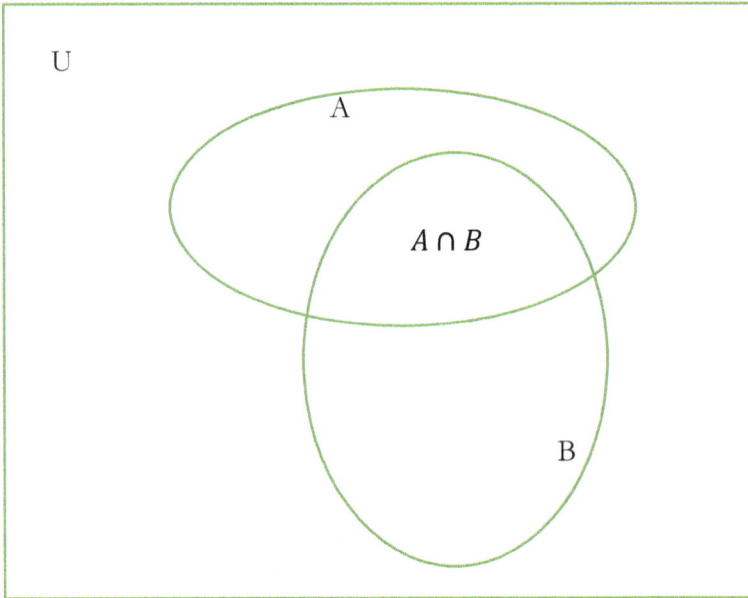

Suppose that 1% of the population have a given disease. Suppose a test for the disease has a sensitivity of 80%, meaning that 80% of people with the disease produce positive test results. This is called the test Sensitivity (or True Positive Rate, TPR). Further suppose that the test is 90% specific, meaning that 90% of people without the disease will produce a negative test result. This is called the test Specificity (or True Negative Rate, TNR). Suppose a randomly chosen person has a positive test result. What is the probability that the person has the disease? Intuition leads most people astray yet again. There is a strong tendency to think that the individual is 80% likely to have the disease. This is wildly wrong, however.

Let A represent having the disease, and let B represent a positive test. We want to calculate $P(A|B)$, the probability of having the disease given a positive test. But we know that the "prior" probability of having the disease, that is the probability before any test is carried out, is $P(A) = 1\%$. We also know that $P(B|A)$, the probability of getting a positive test given that the person has the disease, is 80%.

Finally, what about $P(B)$, the probability of a positive test? This equals the probability of having the disease (1%) times the probability of a person with the disease testing positive (80%) *plus* the probability of not having the disease

(99%) times the probability of a person without the disease testing positive (10%), making 0.01 x 0.8 + 0.99 x 0.1 = 0.107.

We can now use Bayes' Theorem to give,

$$P(A|B) = \frac{P(B|A)P(A)}{P(B)} = \frac{0.8 \times 0.01}{0.107} = 0.075$$

So, the probability of the person having the disease is not 80% but only 7.5%, a very marked difference.

Note that the condition that the person in question is randomly chosen from the population is crucial. Very often medical tests are carried out only when a doctor has identified a risk, perhaps symptom based or due to familial history. Such individuals do not have a $P(A)$ of 1% but some heightened value of $P(A)$ due to being an at-risk sub-population. In practice it would be essential to take that into account.

Particular controversy can arise in the context of evidence before a court of law that may see the defendant sent to prison for a very long time. Suppose that A represents the accused being guilty. Suppose that B represents some piece of evidence pointing towards guilt (say some sort of positive ID such as fingerprint or DNA evidence, or a positive witness ID in a line-up). Let's say that the "sensitivity" of the ID test is 99%, so that guilty parties will be positively identified in 99% of cases. On the other hand the "specificity" of the ID test is 95%, so that innocent parties will be correctly given a negative ID in 95% of cases.

$P(A)$ is the probability that the accused is guilty prior to any evidence being taken into account. For now let's just say there are N people who might feasibly be the guilty party, and prior to the ID evidence there was nothing to choose between them. In other words, $P(A) = 1/N$. What is the probability that the accused is guilty given a positive ID?

$P(B|A)$, the probability of a positive ID assuming guilt is given to be 99%. $P(B)$ is the probability of a positive ID and hence equals the probability of guilt, $P(A) = 1/N$, times the probability of a valid positive ID (99%) plus the probability of innocence, $(1 - 1/N)$, times the probability of an innocent person receiving a false positive ID (5%). So, Bayes gives us,

$$P(A|B) = \frac{P(B|A)P(A)}{P(B)} = \frac{0.99 \times \left(\frac{1}{N}\right)}{0.99 \times \frac{1}{N} + 0.05 \times \left(1 - \frac{1}{N}\right)} = \frac{0.99}{0.99 + 0.05(N-1)}$$

Now for the tricky issue of how many credible suspects there might be. Let's say there are only 2. In that case the suspect is 95.2% likely to be the guilty party. Send him down!

On the other hand, suppose there are 21 people who are equally likely to be the villain before the evidence is taken into account? The suspect is now only 49.7% likely to be guilty. So, not even guilty on balance of evidence let alone "beyond reasonable doubt", which implies acquittal.

But now consider the case such that here is nothing whatever to tie the suspect to the crime other than the ID evidence. The suspect may be unknown to the victim; they may never have met as far as we know. In that case, what is N? In other words, prior to the ID evidence, what is the probability that this person is guilty? Very small indeed. No greater than anyone else who was within a distance capable of carrying it out. So perhaps N might be the entire population of a city, or at least some district of a city. Many thousands, perhaps millions?

In such a case the probability of the accused being guilty, despite the positive ID, is vanishingly small. For $N = 1000$ it is only 1.9%, and even for $N = 100$ it is only 16.7%. He should definitely be acquitted.

I suspect this is a far more common scenario than we would like to admit. And I suspect defendants in such a position are often sent to prison because juries, barristers and judges alike misinterpret the sensitivity or specificity of the ID, be it 99% or 95%, as the probability of guilt. This is so wildly wrong. But worse – some ID evidence – specifically identification by an eyewitness using a line-up – is notoriously unreliable. So the specificity is far worse than assumed in this example.

Appendix J: The Transformation of 4-Vectors

The Lorentz transformation of 4-vectors is specified here in general form, covering both the covariant and contravariant coordinates, and the inverse transformation. Here we consider "active" transformations in which the object (i.e., the vector or particle) is physically rotated or boosted to a higher velocity. This is as opposed to considering the coordinate system (or, equivalently, the observer) to be rotated or boosted, which would have the inverse effect on coordinates. The rotation (5.4.2) is active in this sense, corresponding to rotating a vector anticlockwise in the x, y plane (i.e., a positive rotation by θ about the z-axis of a right-handed coordinate system).

However, the Lorentz boost of (3.7.1), or (3.7.3), were passive transformations in which the dashed coordinates were obtained wrt an observer who was boosted in the x-direction by a velocity u. If we change this to an active transformation in which the dashed coordinates are those which apply after a particle has been physically boosted, then we get instead the inverse transform,

$$x' = \gamma(x + ut) \qquad t' = \gamma(t + ux) \tag{J.1}$$

and the y, z coordinates are unchanged. In (J.1), γ is the relativistic parameter, $\gamma = 1/\sqrt{1 - u^2}$, not to be confused with the Dirac matrices.

Generally, a Lorentz transformation of the covariant components of any 4-vector is given by,

$$v'_\mu = \Lambda_\mu{}^\beta v_\beta \tag{J.2}$$

If the transformation were simply a rotation about the z-axis then the matrix would be essentially (5.4.2), except it would now be a 4 x 4 matrix and (for a pure rotation) the time component would be unchanged. Similarly, for a boost along the x-axis the $\Lambda_\mu{}^\beta$ are given by (J.1). Note that this identifies t, x, y, z as the covariant coordinates, $\{x_\mu\}$. Hence, covariant coordinates of a 4-vector are defined as transforming like the covariant differentials, dx_μ. This gives us therefore,

$$\Lambda_\mu{}^\beta = \frac{\partial x'_\mu}{\partial x_\beta} \tag{J.3}$$

In contrast, the contravariant coordinates of a 4-vector, written v^μ, transform like the differential operators, $\partial_\mu = \dfrac{\partial}{\partial x_\mu}$ for which the chain rule gives, $\dfrac{\partial}{\partial x'_\mu} = \dfrac{\partial x_\alpha}{\partial x'_\mu} \dfrac{\partial}{\partial x_\alpha}$ from which we define,

$$\Lambda^\mu{}_\alpha = \frac{\partial x_\alpha}{\partial x'_\mu} \tag{J.4}$$

which we see provides the inverse matrix to (J.3). The transformation of any 4-vector's contravariant components is then given by,

$$v'^\mu = \Lambda^\mu{}_\alpha v^\alpha \tag{J.5}$$

The scalar product of any two 4-vectors must be just that — a scalar, which means invariant under any Lorentz transformation. The scalar product in this notation is written $a^\mu b_\mu = a_0 b_0 - \bar{a} \cdot \bar{b}$, recalling that this is invariant under all Lorentz transformations. (To ensure this the contravariant and covariant components are defined to be connected by $a_\mu = \eta_{\mu\nu} a^\nu$ where $\eta_{00} = 1, \eta_{0i} = \eta_{i0} = 0$ and $\eta_{ij} = -\delta_{ij}$). The scalar product is invariant as $v'^\mu v'_\mu = \Lambda^\mu{}_\alpha v^\alpha \Lambda_\mu{}^\beta v_\beta = v^\mu v_\mu$ because the two matrices are inverses, i.e.,

$$\Lambda^\mu{}_\alpha \Lambda_\mu{}^\beta = \delta_\alpha{}^\beta \tag{J.6}$$

In the case of rotations, without relative motion, (J.6) reduces to the orthogonality of rotation matrices (i.e., that their inverse equals their transpose). However, the above notation applies to any Lorentz transformation (and boosts are neither orthogonal nor unitary).

Note that the inverse transformation may be obtained by swapping the dashed and undashed coordinates, so that (J.3) gives,

$$(\Lambda^{-1})_\mu{}^\beta = \frac{\partial x_\mu}{\partial x'_\beta} = \Lambda^\beta{}_\mu \tag{J.7}$$

This shows why it was valid, after (5.6.5), to write $a'^\alpha = a^\mu \Lambda_\mu{}^\alpha$ as long as the dashed coordinates are interpreted as relating to the inverse transformation, because then $a'^\alpha = (\Lambda^{-1})^\alpha{}_\mu a^\mu = \Lambda_\mu{}^\alpha a^\mu$.

Appendix K: The Transformation of Dirac 4-Spinors

§5.6 gave the explicit expression for Lorentz transformations of 4-spinors, namely $\psi' = S\psi$ where the 4 x 4 matrix is $S = exp\left\{-\frac{i}{4}w\sigma_{\alpha\beta}\varepsilon_{\mu\nu}^{\alpha\beta}\right\}$ (see after (5.6.4) for the meaning of the terms). This form of expression provides either a rotation about one of the coordinate axes or a boost along one of the coordinate axes. The most general Lorentz transformation can be described by a product of such transforms, as required.

Here we give two explicit examples,

- A boost along the x-axis;
- A rotation about the z-axis.

In both bases we will derive the spinor transform, S, and also the 4-vector transformation implied by (5.6.5), i.e., the $\Lambda_\mu{}^\alpha$ which arises from S due to $S^{-1}\gamma_\mu S \equiv \Lambda_\mu{}^\alpha\gamma_\alpha$.

K.1 Boost along the x-axis

In this case $S = exp\left\{-\frac{i}{4}w\sigma_{\alpha\beta}\varepsilon_{0x}^{\alpha\beta}\right\} = exp\left\{-\frac{i}{2}w\sigma_{01}\right\}$

From $\sigma_{\alpha\beta} = \frac{i}{2}[\gamma_\alpha,\gamma_\beta]$ we get $\sigma_{01} = \frac{i}{2}[\gamma_0,\gamma_1] = i\alpha_i = i\begin{pmatrix} 0 & \sigma_i \\ \sigma_i & 0 \end{pmatrix}$

Hence $\sigma_{01}^2 = -\mathbb{I}$ (meaning -1 times the 4 x 4 unit matrix)

Expanding $exp\left\{-\frac{i}{2}w\sigma_{01}\right\}$ as a power series gives,

$$exp\left\{-\frac{i}{2}w\sigma_{01}\right\} = \mathbb{I} - \frac{i}{2}w\sigma_{01} - \frac{1}{2!}\left(-\frac{i}{2}w\right)^2\mathbb{I} - \frac{1}{3!}\left(-\frac{i}{2}w\right)^3\sigma_{01}\cdots$$

Hence,

$$S = exp\left\{-\frac{i}{2}w\sigma_{01}\right\} = \mathbb{I}cosh\left(\frac{w}{2}\right) - i\sigma_{01}sinh\left(\frac{w}{2}\right) \tag{K.1}$$

This tells us how a Dirac 4-spinor transforms when it is boosted in velocity along the x-axis, via $\psi' = S\psi$. However, we have not yet attached a physical meaning to the boost parameter w. To do so we look at the implied 4-vector transformation, i.e., the $\Lambda_\mu{}^\alpha$ derived from $S^{-1}\gamma_\mu S \equiv \Lambda_\mu{}^\alpha\gamma_\alpha$. Using (K.1),

$$S^{-1}\gamma_\mu S = \left(\mathbb{I}cosh\left(\frac{w}{2}\right) + i\sigma_{01}sinh\left(\frac{w}{2}\right)\right)\gamma_\mu\left(\mathbb{I}cosh\left(\frac{w}{2}\right) - i\sigma_{01}sinh\left(\frac{w}{2}\right)\right)$$

Expanding the brackets and simplifying using the following, readily checked identities,

$$\sigma_{01}\gamma_0\sigma_{01} = \gamma_0 \qquad \sigma_{01}\gamma_x\sigma_{01} = \gamma_x$$
$$\sigma_{01}\gamma_y\sigma_{01} = -\gamma_y \qquad \sigma_{01}\gamma_z\sigma_{01} = -\gamma_z$$

$$\sigma_{01}\gamma_0 = -\gamma_0\sigma_{01} = -i\gamma_x \qquad \sigma_{01}\gamma_x = -\gamma_x\sigma_{01} = -i\gamma_0$$
$$\sigma_{01}\gamma_y - \gamma_y\sigma_{01} = 0 \qquad \sigma_{01}\gamma_z - \gamma_z\sigma_{01} = 0$$

gives,

$$S^{-1}\gamma_0 S = \left(cosh^2\left(\frac{w}{2}\right) + sinh^2\left(\frac{w}{2}\right)\right)\gamma_0 + 2sinh\left(\frac{w}{2}\right)cosh\left(\frac{w}{2}\right)\gamma_x$$

$$S^{-1}\gamma_x S = \left(cosh^2\left(\frac{w}{2}\right) + sinh^2\left(\frac{w}{2}\right)\right)\gamma_x + 2sinh\left(\frac{w}{2}\right)cosh\left(\frac{w}{2}\right)\gamma_0$$

and, $\quad S^{-1}\gamma_y S = \gamma_y \quad$ and $\quad S^{-1}\gamma_z S = \gamma_z$

Using the usual double-argument formulae then gives, replacing the γ_μ by an arbitrary 4-vector,

$$a_0' = a_0 cosh(w) + a_x sinh(w)$$
$$a_x' = a_x cosh(w) + a_0 sinh(w) \tag{K.2}$$
$$a_y' = a_y \qquad a_z' = a_z$$

This reproduces the required Lorentz boost, (I.1), provided we identify

$$cosh(w) = \gamma = 1/\sqrt{1-u^2}$$

(K.2) is then consistent with (J.1) because we may then derive that,

$$sinh(w) = \sqrt{cosh^2(w) - 1} = \sqrt{\frac{1}{1-u^2} - 1} = \sqrt{\frac{u^2}{1-u^2}} = u\gamma$$

And hence the boost parameter is given in terms of the boost velocity by $w = tanh^{-1}(u)$.

K.2 Rotation about the z-axis

In this case $S = exp\left\{-\frac{i}{4}w\sigma_{\alpha\beta}\varepsilon_{xy}^{\alpha\beta}\right\} = exp\left\{-\frac{i}{2}w\sigma_{xy}\right\}$

From $\sigma_{\alpha\beta} = \frac{i}{2}[\gamma_\alpha, \gamma_\beta]$ we get $\sigma_{xy} = \frac{i}{2}[\gamma_x, \gamma_y] = \begin{pmatrix} \sigma_z & 0 \\ 0 & \sigma_z \end{pmatrix}$

Hence $\sigma_{xy}^2 = \mathbb{I}$ (meaning the 4 x 4 unit matrix)

Expanding $exp\left\{-\frac{i}{2}w\sigma_{xy}\right\}$ as a power series gives,

$$exp\left\{-\frac{i}{2}w\sigma_{xy}\right\} = \mathbb{I} - \frac{i}{2}w\sigma_{xy} + \frac{1}{2!}\left(-\frac{i}{2}w\right)^2\mathbb{I} + \frac{1}{3!}\left(-\frac{i}{2}w\right)^3\sigma_{xy} \ldots$$

Hence,

$$S = exp\left\{-\frac{i}{2}w\sigma_{xy}\right\} = \mathbb{I}cos\left(\frac{w}{2}\right) - i\sigma_{xy}sin\left(\frac{w}{2}\right) \tag{K.3}$$

This tells us how a Dirac 4-spinor transforms when it is rotated about the z-axis, via $\psi' = S\psi$. However, we have not yet confirmed that w can be interpreted as the angle of rotation. To do so we look at the implied 4-vector transformation, i.e., the $\Lambda_\mu{}^\alpha$ derived from $S^{-1}\gamma_\mu S \equiv \Lambda_\mu{}^\alpha\gamma_\alpha$. Using (J.3),

$$S^{-1}\gamma_\mu S = \left(\mathbb{I}cos\left(\frac{w}{2}\right) + i\sigma_{xy}sin\left(\frac{w}{2}\right)\right)\gamma_\mu\left(\mathbb{I}cos\left(\frac{w}{2}\right) - i\sigma_{xy}sin\left(\frac{w}{2}\right)\right)$$

Expanding the brackets and simplifying using the following, readily checked identities,

$$\sigma_{xy}\gamma_0\sigma_{xy} = \gamma_0 \qquad \sigma_{xy}\gamma_x\sigma_{xy} = -\gamma_x$$
$$\sigma_{xy}\gamma_y\sigma_{xy} = -\gamma_y \qquad \sigma_{xy}\gamma_z\sigma_{xy} = \gamma_z$$

$$\sigma_{xy}\gamma_0 = \gamma_0\sigma_{xy} = \begin{pmatrix} \sigma_z & 0 \\ 0 & -\sigma_z \end{pmatrix} \quad \sigma_{xy}\gamma_z = \gamma_z\sigma_{xy} = \begin{pmatrix} 0 & 1 \\ -1 & 0 \end{pmatrix}$$
$$\sigma_{xy}\gamma_x - \gamma_x\sigma_{xy} = 2i\gamma_y \qquad \sigma_{xy}\gamma_y - \gamma_y\sigma_{xy} = -2i\gamma_x$$

gives,

$$S^{-1}\gamma_x S = \left(cos^2\left(\frac{w}{2}\right) - sin^2\left(\frac{w}{2}\right)\right)\gamma_x - 2sin\left(\frac{w}{2}\right)cos\left(\frac{w}{2}\right)\gamma_y$$

$$S^{-1}\gamma_y S = \left(cos^2\left(\frac{w}{2}\right) - sin^2\left(\frac{w}{2}\right)\right)\gamma_y + 2sin\left(\frac{w}{2}\right)cos\left(\frac{w}{2}\right)\gamma_x$$

and, $\quad S^{-1}\gamma_0 S = \gamma_0 \quad$ and $\quad S^{-1}\gamma_z S = \gamma_z$

Using the usual double-angle formulae then gives, replacing the γ_μ by an arbitrary 4-vector,

$$a'_x = a_x cos(w) - a_y sin(w)$$
$$a'_y = a_y cos(w) + a_x sin(w) \qquad\qquad (\text{K.4})$$
$$a'_0 = a_0 \qquad\qquad a'_z = a_z$$

which agrees with (5.4.2) and confirms that w is the angle of rotation.

Appendix L: The Application of Biquaternions to Relativity and Maxwell's Equations

This Appendix is based on Lambek (2013). Here we show that, using biquaternions, the general Lorentz transformation of a 4-vector can be expressed in a form which looks almost the same as the quaternionic expression of rotations, namely $p' = qpq^{*\dagger}$, and also that the complete set of Maxwell's equations can be expressed as just one equation, $DF + J = 0$. In addition, the Lorentz transformation of the electromagnetic field, F, in biquaternion form is also very simple: $F \rightarrow F' = q^{*}Fq^{*\dagger}$.

L.1 Biquaternions

The general quaternion is $w + xI + yJ + zK$ where w, x, y, z are real. The general biquaternion is $q = w + xI + yJ + zK$ where w, x, y, z may be complex.

The dagger notation denotes the quaternion-conjugate, such that,

$$q^{\dagger} = w - xI - yJ - zK$$

noting that this leaves the (generally complex) coordinates, w, x, y, z, unchanged. In contrast, the complex-conjugate is, as normal,

$$q^{*} = w^{*} + x^{*}I + y^{*}J + z^{*}K$$

Note that, for any two biquaternions, $(pq)^{\dagger} = q^{\dagger}p^{\dagger}$, which corresponds to the similar expression for the Hermitian conjugate of complex matrices. In contrast, $(pq)^{*} = p^{*}q^{*}$.

When dealing with biquaternions, the terms "real" and "imaginary" become ambiguous. I use these terms to relate solely to the complex context. For the quaternionic part I refer to w as the "scalar" or "temporal" part (whether it is real or complex), and the $xI + yJ + zK$ part as the "vector" or "spatial" part, whether the coefficients are real or complex. Hence, the term "the real part of q" would mean $\mathcal{R}(w) + \mathcal{R}(x)I + \mathcal{R}(y)J + \mathcal{R}(z)K$. In contrast the scalar part is denoted $\mathbb{S}(q) = w$ and the vector part is denoted $\mathbb{V}(q) = xI + yJ + zK$, both of which might be complex.

Whilst the quaternions form a division ring, as every non-zero quaternion has an inverse, the biquaternions do not form a division ring as not every non-zero element has an inverse (for example, $1 + iI$ has no inverse). For non-zero quaternions there is a positive definitive norm whose square is defined by $N(q) = qq^\dagger = w^2 + x^2 + y^2 + z^2$ so that $q^{-1} = q^\dagger/N(q)$. In contrast, for biquaternions, $qq^\dagger$ will not generally be real, or positive definite even when it is real, whilst $qq^{*\dagger}$ is not generally scalar. The lack of an inverse for the general biquaternion is essential for its utility in relativity since it relates directly to the Minkowski metric not being positive definite.

L.2 The Lorentz Transformation

Spacetime points (events) are denoted by special biquaternions, referred to as Hermitian biquaternions, defined as having a real scalar part and a purely imaginary vector part. Thus, if t is a real time coordinate and $\bar{r}$ is a real 3-vector, which in quaternion form is written $\bar{r} = xI + yJ + zK$, where x, y, z are real, then the event is represented by $q = t + i\bar{r}$.

Hence, Hermitian biquaternions are defined such that they equal their full conjugate, $q^{*\dagger} = q$ (equivalently $q^* = q^\dagger$). The squared-norm of an Hermitian biquaternion is thus,

$$N(q) = qq^\dagger = (t + i\bar{r})(t - i\bar{r}) = t^2 - [x^2 + y^2 + z^2] \qquad (L.2.1)$$

where, despite appearances, the minus sign characteristic of the Minkowski metric, occurs because of $I^2 = J^2 = K^2 = -1$. This immediately links to the Lorentz transformation which can be defined as the most general transformation which preserves the norm of an Hermitian biquaternion.

It follows that the transformation of an arbitrary Hermitian biquaternion, p, defined by $p \to p' = qpq^{*\dagger}$ where q is any biquaternion with unit norm (i.e., with $qq^\dagger = 1$) is a Lorentz transformation. This is because,

$$p'p'^\dagger = qpq^{*\dagger}(qpq^{*\dagger})^\dagger = qpq^{*\dagger}q^*p^\dagger q^\dagger = qpp^\dagger q^\dagger = pp^\dagger qq^\dagger = pp^\dagger$$

thus showing that the norm (which is the Lorentz scalar product for an Hermitian biquaternion) is invariant, as required. Note we have used the fact that $qq^\dagger = 1$ implies $q^{*\dagger}q^* = 1$ and that $pp^\dagger$ is a scalar.

If we write $q = u + iv$, where u and v are (real) quaternions, then the requirement for q to represent a Lorentz transformation, i.e., $qq^\dagger = 1$, becomes $uu^\dagger - vv^\dagger = 1$ and $uv^\dagger + vu^\dagger = 0$.

q is a rotation if it is a quaternion (i.e., real, hence $v = 0$). We already know that any rotation can be expressed in this way from §5.4. That a spacetime event is now expressed as an Hermitian biquaternion rather than a quaternion does not detract from this as $q(t + i\bar{r})q^\dagger = t + iq\bar{r}q^\dagger$.

q is a boost if it is Hermitian, i.e., if u is purely scalar (temporal) and v is purely vector (spatial), i.e., $\bar{u} = 0$ and $v_0 = 0$. It is reasonable to expect that any boost, i.e., of any magnitude in any direction, can be represented in this way because there are three degrees of freedom remaining after the four variables u_0, v_x, v_y, v_z are subject to the constraint $qq^\dagger = 1$.

In fact, it is readily shown that any Lorentz transformation can be expressed as a rotation followed by a boost by explicit construction. Putting,

$$\mu^2 = uu^\dagger = vv^\dagger + 1$$

it follows that $\mu^2 \geq 1$ because $vv^\dagger \geq 0$ as a real quaternion. Hence μ can be taken as real, positive and non-zero. We can define the biquaternions $r = \mu^{-1}u$ and $s = \mu - i\mu^{-1}uv^\dagger$. Thus, r is a real quaternion and hence represents a rotation, whilst $s^{*\dagger} = \mu + i\mu^{-1}vu^\dagger = \mu - i\mu^{-1}uv^\dagger = s$ so that s is Hermitian and so represents a boost. Finally we have,

$$sr = (\mu - i\mu^{-1}uv^\dagger)\mu^{-1}u = u - i\mu^{-2}uv^\dagger u = u + i\mu^{-2}vu^\dagger u = u + iv$$

So the rotation r followed by the boost s is equivalent to the initial combined Lorentz transformation.

I think you'll agree this is all very much neater than throwing 4 x 4 matrices around, or dealing with things of the form $exp\left\{-\frac{i}{2}wL_{\alpha\beta}\varepsilon_{\mu\nu}^{\alpha\beta}\right\}$.

Note that if an Hermitian biquaternion, b, transforms under q, i.e., $b \rightarrow b' = qbq^{*\dagger}$ then the complex conjugate Hermitian biquaternion, b^*, does not transform under q but under q^*, i.e.,

$$b^* \rightarrow b'^* = (qbq^{*\dagger})^* = q^*b^*q^\dagger$$

(Incidentally, this means that, for every Hermitian biquaternion, it is necessary to state which transformation applies. It is analogous to the difference between covariant and contravariant tensor components).

The product of two Hermitian biquaternions is not Hermitian, just as it is not, in general, for complex matrices, because $(pq)^\dagger = q^\dagger p^\dagger = qp$ which will only be the same as pq if they commute, which generally they will not. Nevertheless, the product of two Hermitian biquaternions has a simple transformation as long as they transform oppositely, i.e., one under q and the other under q^*. To put that differently, if Hermitian biquaternions a and b both transform under q then,

$$ab^* \to a'b^{*\prime} = qaq^{*\dagger}q^*b^*q^\dagger = qab^*q^\dagger \qquad \text{(L.2.2)}$$

This is the simple transformation rule for products ab^* where a and b are both Hermitian biquaternions transforming under q. Note that it is NOT the same as the transformation of an Hermitian biquaternion, which is $p' = qpq^{*\dagger}$.

If we write $a = a_0 + i\bar{a}$ and $b = b_0 + i\bar{b}$ and if the reader notes that the quaternion product $\bar{a}\bar{b} = -\bar{a} \cdot \bar{b} + \bar{a} \times \bar{b}$, where the terms on the RHS refer to the usual 3-vector notation, then we find,

$$ab^* = \left(a_0 b_0 - \bar{a} \cdot \bar{b}\right) + \bar{a} \times \bar{b} + i\left(b_0\bar{a} - a_0\bar{b}\right) \qquad \text{(L.2.3)}$$

The transformation $ab^* \to qab^*q^\dagger$ therefore tells us that the usual Lorentz scalar product, $a_0 b_0 - \bar{a} \cdot \bar{b}$, is in variant under any Lorentz transform, as it should be. The vector part of ab^*, however, is another matter as this changes under both rotations and boosts.

It follows from the complex conjugate of (L.2.2) that, if Hermitian biquaternions a and b both transform under q then,

$$a^*b \to a'^*b' = q^*a^*bq^{*\dagger} \qquad \text{(L.2.4)}$$

L.3 Maxwell's Equations

I shall consider here only the free space equations, though with source terms (charge density ρ, and current flux density $\bar{J}$). To avoid clutter I shall use units in which $c = \varepsilon_0 = \mu_0 = 1$. (Do not be disconcerted by factors of 4π which you may find in other texts – this too is a matter of units).

Initially I make a fairly lengthy presentation of Maxwells equations and transformation properties in vector, then tensor, form. The length of these will stand in contrast to their highly compact expression in terms of biquaternions.

In vector notation, as made popular initially by Heavyside and still used predominantly in undergraduate physics courses, Maxwell's equations are,

$$\bar{\nabla} \times \bar{E} = -\frac{\partial \bar{B}}{\partial t} \qquad\qquad \bar{\nabla} \cdot \bar{B} = 0 \qquad\qquad \text{(L.3.1)}$$

$$\bar{\nabla} \times \bar{B} = \frac{\partial \bar{E}}{\partial t} + \bar{J} \qquad\qquad \bar{\nabla} \cdot \bar{E} = \rho \qquad\qquad \text{(L.3.2)}$$

The electric and magnetic vector fields, $\bar{E}$ and $\bar{B}$, are just that – vectors. But only in the 3-vector sense that they transform as spatial vectors under rotations. The shortcomings of this standard notation have been evident since Einstein's seminal 1905 paper on relativity because these electromagnetic fields do not transform as 4-vectors under the Lorentz boosts. (How could they, they have no fourth component?). Instead, they transform as follows – where the dashed components are the fields seen by an observer moving at speed u in the positive x-direction with respect to the undashed fields.

$$E'_x = E_x \qquad\qquad\qquad B'_x = B_x$$
$$E'_y = \gamma\left(E_y - uB_z\right) \qquad B'_y = \gamma\left(B_y + uE_z\right) \qquad \text{(L.3.3)}$$
$$E'_z = \gamma\left(E_z + uB_y\right) \qquad B'_z = \gamma\left(B_z - uE_y\right)$$

where $\gamma = 1/\sqrt{1 - u^2}$. The burden of Einstein's 1905 paper is that the transformation rules, (L.3.3), are what is required to achieve his postulated "principle of relativity", namely that the Maxwell equations, (L.3.1,2) take the same form for an observer in uniform motion. Account must then be taken of the moving observer experiencing different spacetime coordinates, and difference source terms. The source terms jointly form a 4-vector, $(\rho, \bar{J})$, as does the differential operator $(\partial_t, \bar{\nabla})$, albeit the latter transforms under the inverse Lorentz transformation (contravariantly). For relative motion of the dashed observed at speed u in the x-direction the transforms are,

$$x' = \gamma(x - ut) \qquad\qquad t' = \gamma(t - ux) \qquad\qquad \text{(L.3.4)}$$
$$J'_x = \gamma(J_x - u\rho) \qquad\qquad \rho' = \gamma(\rho - uJ_x) \qquad\qquad \text{(L.3.5)}$$
$$\partial'_x = \gamma(\partial_x + u\partial_t) \qquad\qquad \partial'_t = \gamma(\partial_t + u\partial_x) \qquad\qquad \text{(L.3.6)}$$

Note that (L.3.6) follows from (L.3.4) using the chain rule of differentiation. (In the language of tensors, if (L.3.4) relates to the covariant coordinates, then (L.3.6) are necessarily contravariant components). Using (L.3.3-6) the Maxwell equations seen by the (notionally) stationary observer transform to the identical form of equations seen by the moving observer, namely that the complete set of (L3.1-6) give us,

$$\bar{\nabla}' \times \bar{E}' = -\frac{\partial \bar{B}'}{\partial t'} \qquad \bar{\nabla}' \cdot \bar{B}' = 0 \qquad (L.3.7)$$

$$\bar{\nabla}' \times \bar{B}' = \frac{\partial \bar{E}'}{\partial t'} + \bar{J}' \qquad \bar{\nabla}' \cdot \bar{E}' = \rho' \qquad (L.3.8)$$

Physically, this means that we have no means of telling, from the behaviour of the electromagnetic fields, which observer is "really" moving. This was, of course, Einstein's whole point.

I have spelt out the transformations in vector/component form to illustrate the contrast with the more compact notations that follow. By 1916, with his publication of the general theory of relativity, Einstein had moved to the use of tensor notation – essential when addressing general coordinate transformations. This provides greater compactness, Maxwell's four equations, (L.3.1-2), becoming two tensor equations of considerably greater elegance, namely,

$$\partial_\alpha F^{\alpha\beta} = J^\beta \text{ and } \partial_\alpha F^{\#\alpha\beta} = 0 \qquad (L.3.9)$$

The first of these is equivalent to (L.3.2) whilst the second is equivalent to (L.3.1). The notation reveals what is, these days, taken to be the correct understanding of the electromagnetic field: it is not a pair of vector fields but a single tensor field. Specifically, it is an antisymmetric rank-two tensor, meaning that $F^{\alpha\beta} = -F^{\beta\alpha}$. In the more sophisticated, coordinate-free language of differentiable manifolds, Maxwell's equations (L.3.9) can be written as $\mathbf{dF^\#} = \mathbf{J^\#}$ and $\mathbf{dF} = 0$ where $\mathbf{d}$ is the exterior derivative. In this language, the electromagnetic field, $\mathbf{F}$, is identified as a two-form (which is just a coordinate-free way of saying it is an anti-symmetric rank-two tensor under general coordinate transformations).

The hash on $F^\#$ denotes taking the dual. (An asterisk is normally used, but I avoid that to save confusion with the complex conjugate). In coordinates this is defined by $F^{\#\alpha\beta} = \frac{1}{2}\varepsilon^{\alpha\beta\mu\sigma} F_{\mu\sigma}$ (summed), where $\varepsilon^{\alpha\beta\mu\sigma}$ is the alternating tensor, which equals 1 when its indices are a positive permutation of t, x, y, z, but -1 when a negative permutation , and otherwise zero. In terms of the familiar fields, the tensor components are,

$$\{F^{\alpha\beta}\} = \begin{pmatrix} 0 & -E_x & -E_y & -E_z \\ E_x & 0 & -B_z & B_y \\ E_y & B_z & 0 & -B_x \\ E_z & -B_y & B_x & 0 \end{pmatrix} \qquad (L.3.10)$$

$$\{F^{\#\alpha\beta}\} = \begin{pmatrix} 0 & -B_x & -B_y & -B_z \\ B_x & 0 & E_z & -E_y \\ B_y & -E_z & 0 & E_x \\ B_z & E_y & -E_x & 0 \end{pmatrix} \qquad \text{(L.3.11)}$$

From these, you may readily check that (L.3.9) are equivalent to (L.3.1,2).

You will note that the operation of taking the dual is equivalent to the replacements $\bar{E} \to \bar{B}, \bar{B} \to -\bar{E}$. Consequently, applying the dual operation twice just changes the sign, i.e., $F^{\#\#} = -F$, in other words "dual" is yet another sort of square root of -1.

The Lorentz boost along the x-axis, (L.3.4-6), is given in this 4 x 4 matrix form by the transformation matrix,

$$\Lambda = \begin{pmatrix} \gamma & -u\gamma & 0 & 0 \\ -u\gamma & \gamma & 0 & 0 \\ 0 & 0 & 1 & 0 \\ 0 & 0 & 0 & 1 \end{pmatrix} \qquad \text{(L.3.12)}$$

Being second rank tensors, F and $F^{\#}$ transform under any Lorentz transformation as $F \to F' = \Lambda F \tilde{\Lambda}$ and $F^{\#} \to F^{\#\prime} = \Lambda F^{\#} \tilde{\Lambda}$ where F and $F^{\#}$ are here understood to be represented by the matrices (L.3.10,11). The Maxwell equations then respect Einstein's condition of being form invariant, so become $\partial'_\alpha F'^{\alpha\beta} = J'^\beta$ and $\partial'_\alpha F^{\#\prime\alpha\beta} = 0$. The reader can also check that the above transformations of F and $F^{\#}$ reproduce (L.3.3).

Finally we are ready to express the Maxwell equations in biquaternion form. Define the biquaternion derivative operator by,

$$D = \partial_t - i\bar{\nabla} \qquad \text{(L.3.13)}$$

Note that it is Hermitian. The 4-vector potential $\{A_\mu\} = (\phi, \bar{A})$ is written is biquaternion form as,

$$A = \phi + i\bar{A} \qquad \text{(L.3.14)}$$

noting that it too is Hermitian. In the above definitions it is, of course, to be understood that $\bar{\nabla} \equiv I\partial_x + J\partial_y + K\partial_z$ and $\bar{A} \equiv IA_x + JA_y + KA_z$. Because D and A are both Hermitian, and both transform under q, (L.2.4) gives the following Lorentz transformation,

$$D^*A \to D'^*A' = q^*D^*Aq^{*\dagger}$$

But using the conjugate of (L.2.3) we have,

$$D^*A = (\partial_t\phi + \bar{\nabla} \cdot A) - \bar{\nabla} \times \bar{A} + i(\partial_t\bar{A} + \bar{\nabla}\phi) \qquad \text{(L.3.15)}$$

We recognise $\bar{B} = \bar{\nabla} \times \bar{A}$ and $\bar{E} = -(\partial_t \bar{A} + \bar{\nabla}\phi)$. Moreover, we know from standard theory that we are free to set a gauge for the potentials, and the Lorentz invariant gauge is $\partial_t \phi + \bar{\nabla} \cdot A = 0$. Hence, if we define the biquaternion electromagnetic field by,

$$F = \bar{B} + i\bar{E} \qquad\qquad (\text{L.3.16})$$

then (L.3.15) becomes,

$$D^*A = -F \qquad\qquad (\text{L.3.17})$$

This is merely the biquaternion form of the fields in terms of the potentials. However, (L.3.17) also tells us that the biquaternion field F must transform in the same was as D^*A, and so, from (L.2.4),

$$F \to F' = q^*Fq^{*\dagger} \qquad\qquad (\text{L.3.18})$$

This is a great deal simpler than either the vector or the tensor formulations of the general Lorentz transformation. Moreover, in terms of the biquaternion source $J = \rho + i\bar{J}$, all four of the Maxwell equations, (L.3.1,2), can be written in the highly compact form,

$$DF + J = 0 \qquad\qquad (\text{L.3.19})$$

I leave this as a simple exercise for the reader to check. It is a matter of expanding the RHS of $DF + J = (\partial_t - i\bar{\nabla})(\bar{B} + i\bar{E}) + \rho + i\bar{J}$. Biquaternion equations imply four separate equations, according to the four combinations of real or imaginary parts, and scalar or vector parts. Each of these four corresponds to one of the Maxwell equations in the form (L.3.1,2).

Whether the notational compactness provided by biquaternions has more substantive implications than the merely cosmetic appears to have been answered in the negative by posterity, since it never caught on and was overtaken in mathematical circles by the coordinate-free language of differentiable manifolds, forms and the external derivative, etc. Nevertheless, (bi)quaternions retain their perennial fascination for many people, myself included.

L.4 References

Einstein, A. (1905). Zur Elektrodynamik bewegter Korper. (On the Electrodynamics of Moving Bodies). Annalen der Physik, 17 (1905).

Einstein, A. (1916). *Die Grundlage der allgemeinen Relativitatstheorie.* (The Foundation of the General Theory of Relativity). Annalen der Physik, 49 (1916).

Lambek, J. (2013). *In Praise of Quaternions.* CR Math. Rep. Acad. Sci. Canada, 2013, math.mcgill.ca.
https://www.math.mcgill.ca/barr/lambek/pdffiles/Quater2013.pdf

Index

In lieu of an index you may download a searchable pdf of this book at,

http://RickBradford.co.uk/FiveSquareRoots.pdf

Internet links within the pdf will facilitate access to the references.

Any typos or other corrections, please email rawbradford@gmail.com

Feel free to browse the other material on maths, physics, cosmology and other things on my site: http://RickBradford.co.uk

www.ingramcontent.com/pod-product-compliance
Lightning Source LLC
Chambersburg PA
CBHW071209210326
41597CB00016B/1740